TRILOBITE!

TRILOBITE!

*Eyewitness
to
Evolution*

BY

Richard Fortey

ALFRED A. KNOPF   *New York*   2000

THIS IS A BORZOI BOOK
PUBLISHED BY ALFRED A. KNOPF

Copyright © 2000 by Richard Fortey

www.aaknopf.com

Originally published in Great Britain by HarperCollins
Publishers, London.

Library of Congress Cataloging-in-Publication Data
Fortey, Richard A.
Trilobite! : eyewitness to evolution / by Richard Fortey. —
1st American ed.
p.   cm.
Includes index.
ISBN 0-375-40625-5
1. Trilobites.   I. Title.
QE821.F67 2000
565'.39—dc21   00-034908

Manufactured in the United States of America

First American Edition

For my mother

Contents

Illustrations

Illustrations

Plates

xi

Plates

References in the text to figures refer to images reproduced in the plate sections.

Preface

I have been enthralled by trilobites for more than thirty years. This book is both my homage to them, and an attempt to convey to others something of the pleasure that their study has given to me. In the process, something of the scientific method may be revealed. My last book was a biography of all life, from bacterium to mankind, in which the trilobites were passed over in a page or two. Now I have the chance to turn the focus the other way and to allow my favourite animals to tell their stories in the detail they deserve. Even so, I am as conscious of what I have had to leave out as of what I have been able to include. History can never be told completely, and three hundred million years of history is bound to be more of a précis than a narrative. I wish to persuade the reader of the excitement of recreating vanished worlds, and of seeing ancient seas through the eyes of the trilobites. This is not an academic study, rather, it is an incitement to discovery.

London, October 1999

TRILOBITE!

I

Discovery

Out of season, the bar of the Cobweb Inn at Boscastle is everything a pub should be. There is a low, heavily-beamed ceiling
hung with antique bottles, and a plain floor which is a jigsaw
of flagstones. Photographs of the local women's darts team
hang on the wall, alongside framed, faded newspaper cuttings which record in print the several virtues of the inn. A log
fire gives out rather more heat than is needed. There is no
music save the low buzz of rich vernacular; in November, no
Londoner ventures to the North Cornwall coast. The Cobweb
is a slightly scruffy, comfortable old place, where you can talk
if you need to, but if you feel like saying nothing you can just
watch the flames in the hearth, and nobody will think you
odd if a smile plays on your lips. It takes an effort of will to
leave the dark, comfortable, nourishing womb of the inn, and
emerge, blinking, into the bright world outside; but leave I
must, because I have to find Beeny Cliff before the light fades.
It can be dangerous out on the cliffs after nightfall.

Boscastle is tucked into a cleft on the wild northern coast of
the long peninsula that completes south-west England, and it
is built around a narrow harbour where the River Valency cuts
down to the sea. It is an ancient place, where the cosmetics
of the tourist trade—Witchcraft Museum and knick-knack

shops—have not quite succeeded in smothering a character that was born of slate and hardship. At one time the town comprised almost nothing *but* inns serving miners and sea-men, of which the Cobweb is a survivor, and you can still imagine a dozen different signs advertising their wares all along the crooked street that leads to the haven. The houses are former inns, prettified with features that fail to disguise their boozy origins. The rough local stone gives the buildings their character. Even the Witchcraft Museum is a cottage with an ancient roof that sags crazily under the weight of Cornish slates. On this day the harbour is almost deserted, and I can imagine the place as it must have looked when the poet and novelist Thomas Hardy visited it as a young man, more than a century ago.

I leave the town on the northern side of the harbour where the path zig-zags up the side of the steep valley. There are gorse bushes which even at this time of year cheerfully wave sprigs of yellow pea flowers. Small birds secretively flit across the path—a wren and some stonechats—as if inviting me onwards. From up here I can see piers guarding the long, nar-row harbour entrance, barriers that were already ancient when the first Elizabeth was on the throne. A cold breeze makes me wish I had put on an extra sweater, but I have luck-ily caught an interval between showers. Suddenly, I climb high enough to see the sea. This is one of those days when the furthest horizon is obscured in mist, as if the sea went on for ever. It is not stormy weather, but I can hear the growl of the surf smashing against cliffs, which weave in and out to the south, one after another, sheer to the sea. A white surf-line marks the junction:

> *With its long sea lashings*
> *And cliff side clashings*

as Hardy described this coast. The cliffs are dark, almost black, while the sea is strangely heavy, wrinkled like a pachyderm, so that only the lazily shifting white line of breakers serves to animate the prospect. The town in its secret valley has quite slipped from view; the solitude is absolute. I shelter from the breeze behind a wall, which is overgrown with rounded tussocks of sea campion and thrift. It is constructed mostly from blocks of slate; curiously, the slate slabs are placed vertically, so that they look like books set on their edges, pages towards you. I am accustomed to different, horizontally-built stone walls around Oxford. The pattern is broken by occasional piers incorporating angular blocks of white, coarse-looking vein quartz. The artisans who built these walls knew their rocks. Slates stacked vertical will let the rainwater (and there is plenty in Cornwall) drain rapidly away, parallel to the way the rock naturally splits. Rubbly quartz is indifferent to all weathers and makes for obstinate pillars. Both rock types are now decorated with a leafy, frilly form of green lichen, which softens every stony outline in such a damp climate.

Now that I study the cliffs I can see that they, too, are made of the same black slates. This is why they seem so forbidding, so stern and dark. In places they are beetling (a word which only seems to apply to brows and cliffs) with teetering over-hangs, fissured, and with obviously dangerous crags. These cliffs are a hymn to vertigo: "haggard cliffs, of every ugly alti-tude . . ." I pay careful attention to the narrow and slippery path; there has been a lot of rain recently, and one foolish step might have serious consequences. Tumbledown stone walls indicate that fields formerly extended very close to the top of the sheer edge, but now there is only a steep grassy slope between the walker and the airy heights where razorbills and fulmars wheel on the wind. The few, stunted trees on the slopes lean away from the fall as if their branches stretched in horror from the tumbling edge.

By the time I reach the top of Pentargon Bay I have some feeling for the geology. The dark rocks displayed on the inac-cessible cliffs have surely suffered in a great vice of Earth movements, for they are tilted and crimped. No strata follow a straight line, instead they take off on a convoluted journey of their own. On the far side of the bay I can see a fissure that extends vertically from cliff edge to sea, which has been excavated by the elements over millennia. This is certainly a fault—a great fracture through the black rocks—a dislocation which must have once made the Earth shudder and tremble. Faults are the visible signatures of earthquakes, sealed for eternity in the rocks. The whole coast must have been gripped by a mighty upheaval causing the strata to crack and buckle. The evidence of a prehistoric paroxysm of the crust is imprinted on these heights.

Look harder, and evidence of tectonism is everywhere. Not far from the fault a stream follows a narrow valley which has been excavated along another plane of weakness in the rocks. Where the stream reaches the sea its valley is cut off abruptly at the cliff, and the brook suddenly plunges into a waterfall two hundred feet above the sea, where it is whipped up by the breeze into spray. Near the water-line there are caves and smaller crevices which have been excavated by the probing sea. Even on such a calm day I can hear the suck and cough of waves assaulting the slates, picking out the weakest spots where folds have cracked the strata, marking each small fault with a chasm or a hole. From time to time a wave rushes into a cave compressing the air within it—which then recoils with a report. It makes a sound like distant cannon fire, an irregular salvo fired in an orogenic war. Imagine the battery on a stormy day. Now it is possible to comprehend how thousands of years of erosion have eventually isolated stacks and islands, like Meachard off Boscastle harbour. In time, these outposts of land will be worn quite away and returned to the sea. I can identify white quartz in the matrix of the gloomy cliffs, as clearly as scribblings of chalk on a blackboard. There is even a patch where the quartz trace shows the strata to have been folded over completely—turned upside down. I can only speculate on the massive forces which have treated solid rock with such disdain. Thicker masses of quartz are aligned along the faults. Squeezed from the rock like serum from a wound, it congeals in the cracks. This must have been the source of those large lumps in the stone walls. Elsewhere, it fills in voids in stressed rock like some kind of mad spaghetti. Ultimately, though, quartz is tougher than slate, and survives as pebbles long after the country rock has been eroded away. I would be willing to wager that some of the rounded pebbles on inaccessible Pentargon beach far below me are made of quartz. They will outlast these cliffs, and—who knows?—maybe they will outlast the human species.

The sooty shales and slates were once soft muds—sediments—which accumulated deep beneath the sea. Time has transformed them: hardened them, elevated them hundreds of feet above present sea level, and folded them. But how much time?

Where I am standing now, close to the edge, there is a notice in red letters: *Caution! Cliffs are liable to cracking. Take extra care.* And it's true. A stack of shale is teetering outwards into the void. It is hard to escape a shiver of apprehension as you imagine the block tumbling over and over to smash to pieces far below. The next haven up the coast is called Crackington, the name encapsulating the precariousness induced by erosion.

I have a geological memoir for the Boscastle area tucked into my jacket pocket. From the geological map which sketches out the pattern of the outcrops of the rocks I can see something of the tectonic agony which is so patent in the cliffs: rock formations twist and turn over the mapped ground, which is criss-crossed by faults. I can identify exactly where I am standing, on the outcrop of the Boscastle Formation; in the dry language of science the slates are described as early Carboniferous (what in the US would be called Mississippian). This corner of the world has an extremely ancient origin, older than mammals, older even than dinosaurs. These black slates would already have carried their contorted signature as a guarantee of antiquity when *Tyrannosaurus* was king of the hill. When they were first laid down there were only tree ferns and cockroaches and cumbrous amphibians on land. Can there be a better place to reflect upon the vastness of geological time?

The erosion which I can both see and hear is ineffably slow. I could stand here all my life and notice little difference to the cliffs. Maybe a chasm excavated along a fault might seem subtly darker as its girth increased after an exceptional storm. Perhaps a rock fall would leave a scar cleared of campion and grass. But I am certain that when Thomas Hardy stood on this spot he would have gazed upon a comparable scene; my eyes now see what his once saw. To be sure, the vegetation would have changed, but the geological signature of the cliffs would have been legible in much the same way. How can we conceive of the time needed to wear away these cliffs to nothing, to convert all the massed slates into fine silt, quartz veins into pebbles—at first angular, then worn by the constant shuffling of the sea rounder and rounder, until they acquire the contours and colours of a hen's egg? Millennia are irrelevant, species come and go, and still the cliffs stand obstinate against the inroads of time. Yet given *enough* time even this rampart that seems to stand so unflinchingly against the surf will be reduced to nothing, and the flagstones on the floor of the Cobweb Inn will return to sediment, joining all the other works of Man, committed once again to the great cycle of change. Rocks are eroded to sediment: sediment is hardened to rock; rock is elevated above sea level by movements of the Earth, transformed by tectonics; and, thus raised, is once more subject to the assault of the elements. This is the great wheel of the Earth. If Gustav Mahler had taken the geological view, *Das Lied von der Erde* (*The Song of the Earth*) would have been a cycle of erosion and reconstruction endlessly reiterated, enough to try the patience even of those who admire the most mantric of symphonies.

Cornwall once formed part of a vast mountain chain. It

marked one end of the Hercynides, which snaked through Europe just as the Alps do today in the south. The patent folding of the rocks was the result of the slates being trapped in a great tectonic vice that showed no mercy. Rocks buckled, in an attempt to accommodate forces that were irresistible. Every tiny ruckle in the wall of Pentargon Bay is the legacy of suffering under a rule of tectonics so mighty that no mere rock could stand against the imperative of crustal stress. When the rocks were folded, structure was piled upon structure until mountains resulted. Clever geologists from the University of Exeter, like E. B. Selwood, have spent years trying to unpick the buckling. They have interpreted this stretch of coast not as merely folded, but as divided into great slices of crust which have slid over and past one another. Squeezed rocks could not absorb the forces by contortion alone, and were compelled to fracture. To attain equilibrium, vast broken slabs of rock larger than a parish slid away from the centre of the forces at a low angle, like the twisted thorn branches leaning away from the reach of the sea wind. Under the sole of these sliding masses weaker rocks were folded over and over, crumpled like a pack of playing cards in the hands of a ruined gambler. Every crack that was opened up, as the country was ground beneath the tectonic wheel, would be filled with vein quartz. Now, the eroded remains of these mountains lay before me. The lichen-covered stone walls were fabricated from the relics of ancient Alps. The farmer who orientated the slaty slabs with such care was conniving with tectonic forces of which he probably knew nothing.

Not many miles to the south, near Bodmin, a granite tor rises above the general plain. It is reminiscent of some stepped Mayan pyramid, not least in scale, but is entirely natural. The strange pile of enormous blocks is what remains behind when granite is weathered over many thousands of years. Even granite eventually succumbs to the onslaught of the elements, rain and wind and frost. But granite endures longer than shale, as I was to see in St. Juliot's churchyard, not far away.

Granite, too, is part of the narrative of the vanished mountain chain, although its source was utterly different from the shales of the Cornish cliffs. It was crystallized from a liquid magma, hot and invasive, within the deep heart of the former chain. Look now: you can see big crystals of felspar, and maybe the sparkle of mica. These crystals tell the story of how a mountain chain drives crumpled rocks so deep into the Earth's crust that they melt, and brew a buoyant, hot broth of liquid minerals, which once more rises through the crust to crystallize and solidify as granite batholiths and plutons. Granite lies at depth beneath the Cornish peninsula, reaching to the surface under the boggy stretches of Dartmoor and Bodmin.

At the moment of their formation, some of the crystals set in motion radioactive clocks. Precise modern instruments can now assess the ticking of geological time as it is recorded by decaying radioactive isotopes of uranium or potassium (and several other elements besides) contained in the substance of the crystals. This method provides the answer to that difficult question: how much time has passed? The decay rate is known: it is only a matter of exact measurement and careful calculation to obtain the date of the mineral's formation. If the Bodmin granites were intruded *into* the folded rocks, then it follows that the granites must postdate the folding. The crystals act almost like eyes that let us see into the past, to calibrate it, to fix it in our vision. So if the crystal age of the granite was 300 million years, it fixes our cliffs as older still—the black slates must have already been folded before the granites were intruded.

As to the soft muds which were eventually hardened and distorted into the black slates, they were deposited on a Carboniferous sea-floor, as long as 340 million years ago. Time hardened them, tectonics twisted them, granites punished them with plutonic heat, long before they came to form the cliffs which now make sites for the inaccessible nests of fulmars and kittiwakes. But messages can be passed on from that ancient sea: messages in the shape of fossils. When the sea-floor was young, shells of many kinds of animals lay littered among the muds and sands, just as clam shells can be found stranded on a beach today. These were mostly ordinary, small creatures, snails and brachiopods and the like. Their shells became incorporated into the sediment as more fine mud rained down upon them: mud originated from the erosion of an ancient landmass, which itself was the product of a previous cycle of Earth history. This is how the story of the world turns and turns again. With the passage of time—much more time—the shells were still there as the muds became deeply buried, and then hardened into shales when water was driven out. Perhaps the substance of the shells was augmented by a subtle infiltration of minerals at this time. Their original colours were leached away, a time-change that bleached shells of brightness, converting them to fossil-colour: they became stony simulacra of once-living creatures.

Their journey had only just begun. The Carboniferous sea in which animals once flourished and left their shelly tokens was consumed in the engine of plate tectonics. Rocks and fossils alike—the accumulated legacy of the sea—were passengers on a great journey. Many of the fossils were doomed to obliteration. They might be dragged to the centre of the growing Hercynian chain, fated to be squeezed or baked beyond recognition. They might be dissolved away. They might be chopped into pieces as the rocks carrying them recrystallized. Mountains grew over south-west England, great slabs of country were shrugged sideways in the melée. Granites insinuated themselves into the depths. As soon as the mountains were born, they were destined to die by erosion, so, most likely of all, the fossils could be worn away to an impalpable mash that was already on a journey to the next Earth cycle. We must marvel at the fossil survivors, their *chutzpah* in the face of the orogenic enemy.

Sealed in the aftermath of tectonics, the fossils still had to survive all that followed: the rendering of a mountain chain once more into the sea. Through more than 200 million years the Hercynian relic was ground down to its roots. It is certain that the granites reached the surface at a time when dinosaurs were still lumbering over the Weald and western Europe, for curious and distinctive minerals derived from these granites are found for the first time in rocks laid down in Cretaceous times, about 100 million years ago. Like a geological strip-tease, veils of rock were stripped slowly away to ever more fundamental levels within the ancient chain. Eventually it was stripped naked to its interior, and the show was over. What I was looking at in Pentargon Cliff was one of the inner veils, preserved from obliteration, with its strata still crumpled like discarded chiffon.

What fossils might survive in the dark slates? What miracles of endurance might they record, what cheating of the laws of chance? How could a wanderer along this slippery path high above the everlasting sea truly comprehend what is meant by the vastness of geological time, even though the evidence is everywhere displayed? Looking down on Boscastle I could almost capture the historical past—I could "see" it as one might episodes from half-remembered movies. It is not difficult to paint a mental picture of Hardy on this path. Or to envisage, more than a century ago, filthy slate miners tottering off to the inn of their choice, with the gentry nearby spry in traps; or to conjure up a bustling Tudor port with heavy-rigged ships sheltering from the wildness of the sea in the safe haven, talk of the Armada in the inns, costumes out of Holbein; I can even visualize an Iron Age farmer's tilth and husbandry, and the discomfort of a November day like this in a simple, smoke-filled dwelling. My vision is full of details rooted in shared humanity, set in bric-a-brac drawn from memory plausibly arranged. But to quantify the geological time required for the formation of Cornwall I need to multiply time a thousand-fold, and then perhaps nearly a thousand times again. I am as accustomed to writing figures in millions (of years) as is a Swiss banker (in dollars), yet the lines of zeros do not translate in proportion. Just as the average working man can understand exactly the purchasing power of fifty bucks, and fifty thousand is probably comprehensible, 500,000,000 has an approximate feel to it—surely, it's a fortune, but what does it mean? A lottery win of 5 million is a lot, but then, so is 22 million. We reel away from such vertiginous figures, we can envisage a huge pile of money, notes piled on notes, but we cannot comprehend its true magnitude. If our intention is to see so far into a past of many millions we shall need to develop a special way of seeing, a spyglass trained on former worlds. We will need to cultivate an indifference to magnitude, so that a million years becomes not, after all, such a long stretch of time. We will need to read rocks and cliffs as if they were books, and not shudder at the heights.

I have passed the steep side of Pentargon Cliff. Some kind soul has carved steps to ease the climb, but even so I am out of breath by the time the path levels off. It now follows a course along the middle of a steep, grassy slope, and is very slippery. There is a curious sense of suspension; the sea is far below, but the slope conceals the vast cliffs which I know must be there. I can still hear the sea clearly enough: there is an irregular bang! bang! as waves punish a sea cave along the invisible surf-line, but the great height I have now reached feels almost illusory, as if I were ambling along some unspecified stratum floating between sea and sky. I have reached Beeny Cliff, and I am grateful that the light has not yet started to weaken. A few drops of rain hit me hard in the neck. A flock of gulls suddenly rises up from beyond the cliff edge, buoyed by the updraught and mewling hysterically. The end of my walk makes me turn up my collar, and shiver.

Beeny Cliff is the scene of a terrifying episode in Hardy's novel *A Pair of Blue Eyes*. Stephen Knight follows the same path that I had just completed, in the company of Elfride, the first of Hardy's complex, closely-observed women heroines. Knight is a man of scientific bent. Perhaps seeking to display his knowledge—or maybe just to satisfy his curiosity—he attempts to demonstrate the contrariwise circulation of the air

GEOLOGICAL TIME SCALE FOR TRILOBITES

Era	Period	Age Ma
		250
	Permian	
	-------------------	290
	Carboniferous	
Upper Palaeozoic	-------------------	354
	Devonian	
PALAEOZOIC +++++++	-------------------	417
	Silurian	
	-------------------	443
	Ordovician	
Lower Palaeozoic	-------------------	491
	Cambrian	
	--	545
	Precambrian (Vendian)	

currents up the face of the cliff: "an inverted cascade . . . as perfect as Niagara Falls—but rising instead of falling, and air instead of water." He leaps on to the slope below the path, and his hat is caught by the counter-current; in a foolish attempt to retrieve it he slips down the appalling incline. He ends up dangling desperately at the edge of the cliff-face itself; Hardy describes the black slates with some precision. Here it is that Knight comes face to face with the subject of this book.

By one of those familiar conjunctions of things where-with the inanimate world baits the mind of man when he pauses in moments of suspense, opposite Knight's eyes was an imbedded fossil, standing forth in low relief from the rock. It was a creature with eyes. The eyes, dead and turned to stone, were even now regarding him. It was one of the early crustaceans called Trilobites. Separated by millions of years in their lives, Knight and this underling seem to have met in their place of death. It was the single instance within reach of anything that had been alive and had had a body to save, as he himself had now.

In this bleak spot, where I was gazing out towards the precipice between the wrinkled sea below and the darkening sky above, trilobites made a brief appearance in English literature. The path above Beeny Cliff was where two paths of my own life—trilobites and writing—uniquely intersected. I felt compelled to visit the place; and I was not disappointed. The eyes of the trilobite, "turned to stone," provided the image I needed to guide the reader through this book, which will try to see the world through the eyes of fossils as a means to animate the past. Equally, I was intrigued by the difference between the novelist's truth, which has nothing to do with testability and everything to do with the impact of the work on the mind and emotions, and scientific truth, which has everything to do with testability, but also with the emotions of discovery—many of them the stuff of novels.

So to what extent was Hardy telling the truth? Hardy was writing his novel originally as a serial—he needed to keep his readers enthralled, episode by episode. Knight's predicament was a cliffhanger in the most literal sense. How can you have more suspense than to leave your character exactly suspended? The trilobite eyes provided a focus for Knight's near-nemesis; while human eyes—blue ones—provided both the title and emotional motor for the novel. The critic Pamela Dalziel has noticed how the plot twists upon "spying" of one kind or another. The book is soaked with the implications of sight.

I was interested to see how closely Hardy had observed the setting for his cliff-hanging episode. Scholars have identified the location from many details from his early life. While he was still employed as an architect restoring St. Juliot's Church in 1870, he met his future wife, Emma Gifford, who was the sister-in-law of the Rector. Locations in the area are rather thinly disguised, and it evidently lies well to the west of the rest of Hardy's Wessex. *A Pair of Blue Eyes* included more autobiographical elements than his other novels, and perhaps for that reason evidently remained dear to his heart (he rewrote parts of it many years later). The height of the cliff itself is rather pedantically described, even as poor Knight flounders, surely indicating that the young writer had gazetteers open, and a sharpened pencil in front of him as he catalogued its statistics: "proved by actual measurement not to be a foot less than six hundred and fifty [feet] . . . three times the height of Flamborough, a hundred feet higher than Beachy Head . . . thrice as high as the Lizard" (and there is more in this vein). But what left me in no doubt at all about the site was the walk I had taken, which was duplicated exactly by the characters in the novel. Hardy had observed what was evidently the same waterfall plunging to obscurity that I had seen at Pentargon Bay, "running over the precipice it was dispersed in spray before it was half way down, and falling like rain upon projecting ledges made minute grassy meadows of them." He described the threatening cliffs, with their "horrid personality" at the same point as I had seen them on my walk. In many respects passages in the novel could be described as reportage: he could recognize both quartz and slate, he recounted physical features in order along the path, he knew some geology and meteorology. While Knight clung to the cliff face his mind raced back through geological time—a series of tableaux of the past succeeded one another in his desperate mind all the way back to the ancient time of the trilobites. It is not a bad account, scientifically speaking, of the succession of life through geological time as conceived about 1860.

But now we reach a point at which fiction and description take off along different paths; perhaps the fascination of the novel lies in these deviations. Why did Hardy refer to Beeny Cliff as The Cliff with No Name? The hamlet of Beeny is clearly named on old maps and he was usually adept at near-synonyms—Camelton for Camelford, Dundagel for Tintagel. I believe that "No Name" adds to the horror and mystery of the place. The same point was appreciated by Sergio Leone: in his brooding spaghetti westerns "The Man with No Name," played by Clint Eastwood, was the anti-hero. Naming is the first way we domesticate our surroundings; we would *expect* the highest cliff along the coast to have a name. Anonymity is horror. When a series of murders are committed terror is at its worst while the perpetrator is still unknown: a case of name-less dread. The fabricator of fiction knew this; this is where artifice comes in. Hardy liked facts, but he knew also when to suspend them. The trilobite itself is a convenient fiction. The Carboniferous rocks on this part of the Cornish coast have yielded no trilobites. The rocks are of the right age, so there is no theoretical objection to finding trilobites there. Fossils are very rare from such tectonically tortured rocks. Enough have been found—precursors of ammonites, clams, microscopic fossils—to establish the geological age, but trilobites are not among them. I would be absolutely delighted to receive a trilobite from a collector lucky enough to hammer one out of these unpromising strata; it would be of appreciable scientific importance. Hardy placed the fossil in the right geological set-ting, but the finding of a trilobite is an invention. He *needed* the trilobite to stare out Stephen Knight. As readers of fiction we relish his use of this particular fossil, "but a low type of animal existence," as heightening the drama of the situation. It mat-ters not a jot that the occurrence is fictional, even though the rest of Hardy's account is an almost photographic reproduction of a real scene. A scientist would be appalled if one of his colleagues invented such an occurrence, for science trades on the truth—nothing but the objective fact. The truth of the artist can recombine the facts of the world in the service of creation, but the scientist has a different duty, to discover the truth lying behind the façade of appearance. Both processes may be equally imaginative.

As for Mr. Knight, he escaped his predicament thanks to a rope, but a rope desperately manufactured from Elfride's undergarments. It is a turning point in the novel, symbolizing a change in the relationship between a flawed man and an unusual woman. And we are not compelled to wonder whether such an episode was fact or not, because it is woven into the fabric of a novel.

I left Beeny Cliff by climbing up the steep steps to Fire Beacon Point, a wild bluff commanding views of the whole Hercynian coast. It may have acquired its name as one of numerous beacons on which fires were lit to warn of the approach of an armada. Now, there is only a bench at the top dedicated to the memory of Paul C. Heard, but I thanked Mr. Heard's relatives for the chance to rest and regain my breath. I turned inland, following an old, walled track. I wanted to see the parish church of St. Juliot on which Hardy had worked. It is perfectly set on a sheltered hillside, and behind it a track leads out directly across fields upon which sheep graze unhurriedly. Despite the setting, there is something depressingly foursquare and Victorian about St. Juliot's—possibly the fault of Hardy's restoration—including a rather dumpy tower with unnecessary castellation. It is as if the site deserved something much more distinguished to live up to its associations. After all, Hardy wrote that "much of my life claims the spot as its key." Surely it is an ancient holy place, which the church fails adequately to honour. In the graveyard there are several "Jollows"—possibly a bastardized version of the name Juliot—the kind of family that must have grown out of the local soil as surely as wind-blown gorse. Inside the church there is a notice informing the visitor that most of the church plate has already been stolen. There was nothing to induce me to linger.

But as I left I saw the Celtic crosses, more ancient than the church itself. Close by the gate one crudely shaped column as high as a man stands sentinel; its apex is sculpted into a disc, and you might expect to find some sort of face hewn upon it, but instead there is a schematic cross. Unlike the slatey grave-stones, these upright crosses are made from granite. They will endure when all the inscribed memorials to Jollows have crumbled into the same soil that long since claimed their bones. The granite was derived from one of the igneous intrusions into the bowels of the Hercynian mountain range. Maybe the blocks were derived from Bodmin or Dartmoor, but whatever the source their hewers knew about permanence. Each monument acknowledges the vast span of geological time, the durability of stone compared with a human lifetime. Each is a symbol of the crossing of human intention with tectonic history—the same history that placed Hardy's plausible but fictional trilobite into Beeny Cliff. The discoidal apex of the cross is like an eyepiece with a view clear back to the age of the tree ferns and lungfishes. This, after all, may have been what I wanted to discover on a chilly November day in Boscastle. I felt a curious elation, even as the icy rain-drops began to refresh the lichens that alone could derive sustenance from incorruptible granite—set solid ever since trilobites had patrolled the shallows of a vanished ocean.

If you can have love at first sight, then I fell in love with trilobites at the age of fourteen.

The peninsula of St. David's forms the south-western promontory of South Wales, extending westwards like a miniature version of the Cornish peninsula where Hardy, too, encountered love. Like Cornwall, it is a region of spectacular and ancient cliffs, while inland the scenery is flat and characterless. There are little coves here, too, with names like Solva and Abercastle, formerly wild and remote fishing villages, but now spruce with whitewash over the rough stone. But the cliffs are as wild as ever they were, and displaying rock folds as convoluted as anything in Cornwall. A walk along the coastal path parades one rock formation after another, delineated by contrasts in colour and texture: here massive yellow or purple sandstones plunge like naked ribs into the churning foam, there a group of contorted dark shales zig-zag up the cliffs like some sort of berserk concertina. In Caerfai Bay there are bright red shales, looking improbably pert in a world of dun geology. All of these rocks are still more ancient than their Cornish counterparts. They date from the Cambrian period, the oldest of the shales having been laid down as muds beneath the sea something like 545 million years before the present. This is getting back to the beginning of things, to a time before there were any plants on land, to a time before any kind of backboned animal existed. Yet there were already trilobites to witness this nascent world. These trilobites were 200 million years older than Thomas Hardy's invented fossil (or, I should say, its real equivalents)—a span a hundred times longer than Man's brief tenancy of this planet. This was the time I explored with a coal hammer at a period of my life when my voice had just turned unreliably falsetto and baritone by turns. While others discovered girls, I discovered trilobites.

I had marked the presence of fossils on a local map. They were described as the oldest fossils in the British Isles. What could be more irresistible? There was something extraordinarily exciting about tapping into a vein of such prehistory. The top dressing of the landscape of human tenancy was stripped away to reveal some deeper reality, layer after layer of geological time unpeeled in my imagination. While my long-suffering mother knitted or read, I beat the rocks at Nine Wells and

Porth-y-rhaw.* These were places where the rocks were accessible by foot and could be broken by sheer effort. I did not even have a proper geological hammer. The fever of discovery was upon me. I learned how to break the hard rock so that it split in the same direction as the former sea floor—this way I was more likely to retrieve something recognizable. It was clear that tectonic forces had tipped the strata vertically. I had to scrabble to dislodge reasonable-sized blocks for breakage. I ignored the sharp pieces of gorse that speared the backs of my hands. Time had made the rock both hard and brittle: it seemed to want to break anywhere but in the right direction. On the broken surfaces there were scraps and fragments of what might, or might not have been the remains of past life: black patches, a little shinier than the rest of the rock. Then, at last, I found a trilobite. The rock simply parted around the animal, like some sort of revelation. The truth is that the fossil itself had rendered the rock weaker: it was predisposed to reveal itself, almost as if it desired disclosure. I was left holding two pieces of rock: in my left hand the positive impression of the creature itself (known as the part); in my right hand the negative mould which had once comprised its other half (the counterpart); the two together snuggling up to survive the vicissitudes of millions of years of entombment. There was a brownish stain on the fossil, but to me it was no disfigurement—surely what I held was the textbook come alive. Drawings and photographs could not compare with the joy of actually touching a find which seemed, in the egotistical glow of boyhood, dedicated to yourself alone. This was my first discovery of the animals that would change my life. The long thin eyes of the trilobite regarded me and I returned the gaze. More compelling than any pair of blue eyes, there was a shiver of recognition across 500 million years.

I would one day learn that the trilobite had a name, *Para-*

---

*Both these sites are now legally protected from hammering, although they were not at the time of my schoolboy excursions.

doxides. When we first exchanged glances I knew nothing of classification or nomenclature, and it did not matter to me: there was plenty of time to learn more. What I held was a specimen that fitted comfortably into the palm of my hand. It was clearly divided along its length into three lobes—a convex central portion and to each side of it identical, but slightly flattened, areas. These were the lobes implied by the name—*trilobite*. The whole animal seemed to bulge towards one end. I knew, by some principle which I could not articulate, that the wider end was the head of the animal. And of course upon this head there were the eyes. Despite the unfamiliar conformation of the fossil, I knew that eyes *must* always belong on heads. So despite the exoticism of the fossil there was already a common bond between me and the trilobite—we both had our heads screwed on the right way. I could see that the body was subdivided into a number of little divisions—or segments, as I would learn to call them. Then there were cracks running across the body. These had nothing to do with the original structure of the animal, rather they were testimony to the long journey through geological time that the Cambrian creature had travelled before it fell apart under my hammer blow. They were joints in the fabric of the rock itself, the scars of an adventure that might have seen the trilobite eroded into oblivion or obliterated in the vice of a thousand tectonic accidents.

This book grew out of that first encounter. I want to invest the trilobite with all the glamour of the dinosaur and twice its endurance. I want you to see the world through the eyes of trilobites, to help you to make a journey back through hundreds of millions of years. I will show that Hardy's description of the trilobite as "but a low type of animal existence" was hardly just, but that his placing the animal at the centre of a drama of life and death might have been nearer the mark. This will be an unabashedly trilobito-centric view of the world.

For trilobites have been witnesses to great events. Stephen

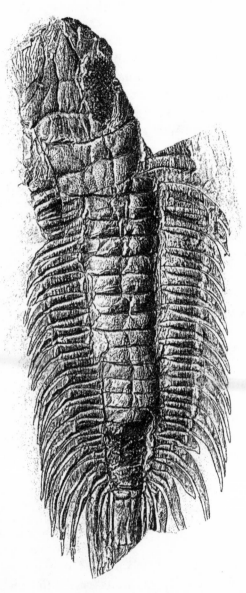

A drawing of the giant trilobite *Paradoxides* by Philip Lake published in 1935, from the same Middle Cambrian rocks in western Wales that yielded the first specimen to my schoolboy hammer. For a photograph of *Paradoxides*, see p. 222.

Knight might have read from the trilobite's stony eyes that the predicament of a mere individual meant nothing. They have seen continents move, mountain chains elevated and eroded to their granite cores, they have survived ice ages and massive volcanic eruptions. No living thing can disengage itself from the biosphere, and trilobites followed the same pattern: their history was also shaped by the events they witnessed. When strangers express their surprise that it is possible to devote a lifetime to studying extinct "bugs" I remind them of how much has happened in the last few thousand years and invite them to imagine what it is to be a historian of dozens of *millions* of years. We are doomed to know so little, like fishermen trying to understand an entire ocean by throwing in a few baited handlines. And if anyone wonders how it is possible to invest such devotion in a group of organisms which died out long ago as a result of who-knows-what inadequacies, there is an obvious answer. Trilobites survived for a total of three hundred million years, almost the whole duration of the Palaeozoic era: who are we johnny-come-latelies to label them as either "primitive" or "unsuccessful"? Men have so far survived half a per cent as long.

There are accounts of scientific research that present the story of discovery as a series of glittering prizes that must be won by the most muscular intellect; this is science as a version of trial-by-combat. Or else scientific research is contained in a metaphor of a journey into uncharted territory, as expressed by Robert Louis Stevenson (in *Pulvis et umbra*): "science carries us into realms of speculation, where there is no habitable city for the mind of man." It is certainly true that there are races to be first in science and that a few massive minds venturing into "realms of speculation" command the most attention—and they deserve it. Such models of scientific progress are typified by mathematicians and physicists, beautifully elaborated by Karl Popper in *Conjectures and Refutations*. Nonetheless, as a description of much of scientific endeavour, both the combative and the adventurous-speculative views are flawed. Many scientists—perhaps most of them—are a curious species for whom the pleasure of finding out is at least as important as the size of the goal. He or she is often a co-operative creature, comfortable with the happy exercise of innate ability, and if a momentous discovery comes it may arrive unexpectedly, like an unanticipated legacy. The unique property of the scientific endeavour is that so many of the regular footsoldiers contribute to the victory. Unlike a poetaster whose burblings are destined for true oblivion while the creations of a Keats survive, even a minor scientist might well make a permanent contribution to a famous campaign—an uncelebrated private who did not die in vain.

Even the most singular fields of scientific inquiry relate in subtle or unexpected ways to larger questions. We shall see that an apparently self-contained and esoteric occupation like the study of trilobites has contributed to mighty debates about the origin of new species, or the nature of major features of evolution, or the distribution of the ancient continents. Those who started with a deep desire to know more about the details of life habits of vanished animals—out of sheer curiosity—may suddenly realize that the detailed knowledge they have accrued relates to something different and more general: something as grand as the structure of an ancient ocean or the arrival of an asteroid on Earth.

I believe that a more accurate image for the way much of science works might be a series of interconnecting paths. Each one has its own interests and delights; sometimes we know where a path leads, on others we are taken by surprise by twists and turns. And where there are intersections with other paths there can be unanticipated new directions which may lead to wholly unexpected views. Like Stephen and Elfride on the path above The Cliff with No Name there may be a crucial conjunction of circumstances which changes everything, and something as small and ancient as a trilobite may be the catalyst for the transformation.

This book will follow a few of the paths that led me from

that first schoolboy find. In pursuit of trilobites I shall visit remarkable places and spend time with remarkable people. Knowledge has been hard won, and there are heroes whose names are known only to me and a few of my friends, who deserve wider recognition. There are stories of personal tragedies which have influenced this tale of trilobites. Discovery isn't a simple matter of "onwards and upwards." It is imbued with all the tawdry and magnificent stuff of human lives. The story of that small part of science which is important to me will illustrate the way this defining human activity works better than some other accounts of greater endeavours: like relativity or the first few nanoseconds of the universe. Sometimes, a miniature gives a better likeness than a grandiose portrait.

Come and see the world as it once was through the crystal eyes of the trilobite. We shall find out how trilobites tell us the pattern of evolution, and how it can be read from the rocks. We shall discover how faith in trilobites not merely moves mountains but shifts whole continents. We shall see how cast-off shells can be re-animated into living animals. We shall understand something of the origins of the richness of the animal kingdom. Through trilobites, we shall take possession of the geological past.

I I

Shells

In 1698 Dr. Lhwyd wrote to his correspondent Martin Lister about the fossils to be found in the limestones around the South Wales town of Llandeilo: "the 15th [of August] whereof we found great plenty must doubtless be referred to the sceleton of some Flat-Fish." Lhwyd's "flatfish" were, of course, trilobites.

When my children were young they used to play a game with sea shells. Holding a large whelk to one ear they contrived to "hear" the sea: the distant crashing of waves on the shore, or the insistent whistling of a gentle sea breeze. Later they understood that the conch merely amplified the murmuring of the air around them, but they never forgot the leap of imagination that joined shell to sea.

Palaeontology is all about listening to what fossil shells have to say. We have to pay attention to shells, because hard skeletons made of durable minerals are almost always what fossilizes. With rare exceptions, the soft anatomy eludes us. Body tissues are food for predators, or for the agents of decay. Who has not picked up a crab shell on a beach, and let it go with a cry of disgust, startled by the pong of putrefaction? Bacteria are everywhere, greedy to decompose, voracious for those organic molecules manufactured with vital energy during life—and then yielding up that energy again to a host of tiny diners, a few thousandths of a millimetre long. "Too, too solid flesh" does indeed melt. What remains behind—shells and bones—has little to sustain the ubiquitous bacteria. Trilobite shells are like those of a dozen or more other kinds of marine animals, being sculpted from the hard mineral calcite. Crab shells are composed of calcite, so are those of clams. If trilobites had not carried their hard shells they would have been effectively invisible to us, for they would have left virtually no trace of their former existence. Had the seas teemed with them as thick as oats in porridge we would have remained ignorant of their rich diversity. Fossil shells are the cast-offs of life, the tough bits, the inedible residue. That which was of least interest to other living animals in life is, ironically, just what generates the interest of both the scholar and the geologist once it has been transformed by time into a fossil. To begin to understand trilobites it is necessary to know about their shells.

Even the shell of a trilobite loses something on death. Colour is the most transient of properties. We know that sea-life today is a symphony of hues: colours flashed as warning, colours subtly employed in disguise, colours seemingly just for sheer exuberance. It is likely that the seas hundreds of millions of years ago were just as polychromic. Yet colour is the first property to bleed away in the fossilization process. The fossil world is a pallid world, which only imagination can revivify. The colours of the dark trilobites I saw in western Wales were the colours of the rocks that entombed them, with no indication of the appearance of the original animals. We can colour them up as we fancy.

I learned trilobite shell anatomy as a student. The terms that I heard empowered me, allowed me to place these strange animals within a field of comprehension. It was curious how learning to call the head of the trilobite the *cephalon* seemed to admit me into the special world of the trilobite lover. *Cephalos* is the Greek for "head" so really we are just replacing one head by another. All trilobites, I learned, were divided into three portions not just along their length—the tri-lobes—but also crossways. As I had instinctively recognized from my first trilobite at St. David's, the cephalon was the front portion, the part that looked at me. At the other end was the tail, which I must learn to call the *pygidium*—a Greek word again.

The adoption of classical language is not to be wondered at, for in the early days of natural history Latin was still the chief medium of communication between scientists of different nationalities. Classics was the *sine qua non* of the educated classes, the *lingua franca*, and not just dropped into any sentence to impress the reader with the learning of the writer (see *sine qua non*, above, but not *lingua franca*, which isn't Latin). Botanists are still obliged to write a shorthand Latin description of every new species (although this may be about to change), something zoologists have not had to worry about for a hundred years. But classical terms for bits of anatomy, whether of an animal or a plant, are more enduring. Medical students curse them even as they commit them to memory, laymen puzzle over them. They serve as a continuous link all the way back to the time of William Harvey and his unscrambling of the circulation of the blood; the terms last while the concepts around them change. This conservatism serves a useful function in retaining a language in which specialists can converse precisely. The first task of the tiro is to master words. The sign that the beginner has joined the hidden club of experts is when he can trade technical descriptions with confidence. There is more to it than that, because the applying the right word is also a guarantee of recognition. The more closely a natural object is anatomized the closer you have to look, and the more terms have to be mastered. Latin or Greek, it matters not, what is important is that a technical term is a shorthand for learning. Nomenclature is a preface to understanding.

So I observed that between cephalon and pygidium there was the *thorax*. It is a familiar enough word, even if its use in the human context is quite different. This portion of the trilobite was the longest part—at least in those trilobites I studied first. It was further divided into a number of segments—thoracic segments. (At the height of Steven Spielberg's fame I toyed with the idea of a movie to restore trilobites to the dramatic centre in the history of life that they deserve. Perhaps a crazed palaeontologist reanimates them from the dead using some appropriate hocus pocus until they rampage unbridled through New York, savaging scantily clad beauties, and knocking down buildings . . . that sort of thing. It was to be called Thoracic Park.)

Each thoracic segment was jointed, and connected by a weak hinge to the one in front and the one behind. Segments form a linked system, rather like railway carriages in a line. They are all more or less similar, and connected together by a coupling. If we had tried to snap the fresh trilobite in two it is almost certain that it would have snapped between segments. It is the same principle that makes us divide a lobster shell at the back of the head. The lack of comparable segments is what makes a tortoise so impregnable—but also so inflexible. A tortoise trundles along, lurching over obstacles, and often dies if it happens to tumble on to its back. There is nothing more ineffectual than a tortoise foolishly waving its legs in the air in a doomed attempt to right itself. Not so a segmented animal. When an obstacle is encountered the segments can shift relative to one another to allow flexibility: they are articulated, hinged, jointed. The movement between segments is dictated by the laws of mechanics: this is why iron plated, alien bugs in science fiction movies look both mechanical and convincing. Segments really *are* all about the articulation of armature. Even when stranded on their backs segmented animals can wriggle upright. To the trilobite a certain vulnerability was the price of flexibility, and a price worth paying. The trilobite

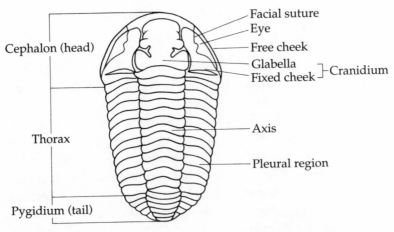

The anatomy of a trilobite: a few technical terms, labelled here on *Calymene*, allow us to describe almost any trilobite.

could move over obstacles, flex and turn, like a train that needed no tracks on which to run.

Looking closer at the tail—or pygidium—it is clear that this, too, included a few segments, but instead of being free to articulate they are fused together, forming a shield. In some trilobites the pygidium is longer than the head, and includes many segments; in others it is minute. Later I would learn why these differences might be useful to the animal. Both thorax and pygidium had an obviously convex central portion— the middle lobe of the trilobite—which, in a rare display of nomenclatural simplicity, is called the *axis*. Furrows divide the axis from the lateral, or *pleural*, parts. So now I had heard the names to distinguish the three lobes of the trilobite: the axial lobe and the flanking pleural lobes. Each thoracic segment had a pleura on either side. In my first ever trilobite these pleurae had spiny tips, so I knew that if I had picked up the living animal it would have felt spiky in my hand, with that particular kind of unpleasant prickliness that one gets with handling a langoustine.

As for the trilobite head that first attracted my attention when I cleaved open the dark Cambrian shale of St. David's, that, too, showed an inflated axial part—the axis of the thorax carried on and widened forwards into a swollen, and prominent, median portion. "This," our professor said, "is the most important and characteristic part of the trilobite—the *glabella*." No familiar ring to this word, it just had to be learned. It had the small advantage of rhyming with "umbrella"—and undergraduates are fond of such *aides-mémoires*, even if they are harder to remember than the original thing to be memorized. The glabella was traversed by furrows that suggested that there was more than one segment in the cephalon, just as in the thorax and pygidium. However, the segments must have been fused together to make the head end an altogether more robust contraption than the thorax, and more like the pygidium in this respect. To either side of the glabella are the eyes, and—believe it or not—they are just known as *eyes*. By using such a simple word we acknowledge our realization of the connection between the student observing and the object studied. Hardy's "eyes, dead and turned to stone" looked across millions of years with a stare of genuine recognition of like for like.

So in just eight technical terms—cephalon, thorax, pygidium, segment, axis, pleura, glabella, eyes—it is possible to begin to embrace the form of these strange animals. To be able to name the parts introduces a certain familiarity. Further, to be competent to recognize the glabella for what it is means that it does not take long to see that one trilobite has a glabella which is quite different from that of another. With language comes discrimination. And it is true that every item named is capable of varying mightily from one species to another: there are those with large eyes and small, those with long thoraces and short, wide or narrow pygidia. I was soon to learn that there were thousands of different species of trilobites and, in the end, I was to become a namer of new names myself.

For now, though, trilobites were hardly more than the jetti-soned husks of once-living animals. Listening to the message of the shells was like trying to hear the sounds of a far-off sea, more remote than childhood. I had begun to acquire a lan-guage to describe what I might be hearing. The reader will need to be equipped with the same short list of terms in order to follow the rest of the tale of the trilobite: it is not so difficult to memorize them. The inventory of trilobite parts that I first learned were the same bits of anatomy that the original dis-coverers of these fossils had recognized as long ago as the eighteenth century. These pioneers were puzzled as well as excited by trilobites: they gave them names like *Agnostus* and *Paradoxides*, which surely reveal the problems they had with their interpretation. There is even a Cambrian species called *Paradoxides paradoxissimus*, which might be translated from the Latin as "the most paradoxical paradox"—more paradoxi-cal than which it is not possible to get.

These early observers soon realized that the shells prised from the rock were only carapaces, rather than the whole animal. The trilobite they knew was only the back of a compli-cated organism. The shell was nothing more or less than a shield which covered its upper side—the part which is most exposed to a hostile world. A shield protects. In old liter-ature it is quite common to find the cephalon referred to as the headshield—and the pygidium as the tailshield—and these descriptions are still entirely appropriate. Tough calcite made the trilobite less vulnerable on its dorsal side, while its under-side, the ventral side, was the sheltered underbelly where lay the soft anatomy that is so rarely preserved. The trilobite seems almost pathetically unprotected on its lower side, for the shell stopped abruptly after being "tucked under" the edge of the trilobite in a narrow shelf, or selvage, known as the *doublure*. Beyond the doublure there is only a cavity, an absence of evidence. This is quite different from the tortoise, where the underside is sealed by a bone shield called the plastron—making it into a veritable tank. The trilobite is half a tank. There is no living creature exactly comparable, although if you turn a woodlouse (or slater) on its back, where their little legs kick and struggle is the equivalent of the trilobite's body underneath. For many years what lay beyond the doublure of the trilobite was a mystery—the trilobite was like a paten without the bread at Eucharist, a vessel lacking its full significance. You will learn how the mystery of legs was solved in the next chapter.

I acquired my first trilobite facts from lecturers and professors. When I was a student it was possible to assimilate most of the basic elements from textbooks, if you wished, and nowadays we can summon massive inventories of information from the web, but it still makes a difference to learn directly from a real scholar. This experience reaches back to a time when the oral tradition was the only way of teaching— when the young received wisdom as a favour from the elders. In China, despite the Cultural Revolution, this reverence for the wisdom of the aged endures. When I was in Nanjing in 1983, I was taken to see the grave of Professor Grabau, a western palaeontologist who almost singlehandedly introduced modern geological principles into China in the early years of the twentieth century: he was, so I was told, "a great teacher." The Chinese compliment is such an important one that it is represented by a special ideogram. It was a simple grave, but obviously regularly and lovingly tended. I was once briefly favoured with the status of "great teacher" myself: I received a letter from a far eastern student who had evidently learned his English from Rudyard Kipling and Rider Haggard. "Oh Great Palaeontologist!" it began, "may I sit at your feet?" Those who know my feet might regard this as unwise, but I was touched by the faith evinced in the oral tradition.

My own guru was Professor Harry B. Whittington. He is the doyen of trilobites, the head—the cephalon—of the tribe. He would teach me to hear the messages of the shells. Somehow, I had managed to convert an infatuation into a career.

I learned my trade on the frozen ground of Spitsbergen beyond the Arctic Circle at 80° North, with the Valhallfonna icecap defining the skyline and an iceberg-clogged sea before me. An amazing variety of new trilobites was discovered in Ordovician (470 million year-old) limestones on the northern part of Spitsbergen, and I was lucky enough to be there when they were found. I collected rock beds exposed along the shore one by one in the appropriate order, so as to turn the stony pages of the trilobite diaries in sequence, tap-tap-tapping my way through my own little piece of geological time (a mere 10 million years or so). Mostly this consisted of bashing hard rocks with a geological hammer, until they were in small pieces on which fragments of trilobite could be seen. Hardened criminals used to be required to do the same thing before it was banned as inhumane. I loved it. All discomfort in the harsh climate was set aside in the inspiriting warmth of discovery. You never knew what the next hammer blow might bring, and occasionally there was something astonishing. The collections were arranged logically, the oldest specimens from the lowest rock beds exposed were carefully labelled and wrapped first, then progressively younger collections, and all were shipped back to the Sedgwick Museum in Cambridge, to which I eventually returned.

I lived in the Sedgwick Museum for nearly three years, happy as Larry. At that time research students shared offices in the attic of the dowdy museum, a nineteenth-century mock-Gothic building on Downing Street, which still houses the Department of Earth Sciences. I drove my room-mate John Bursnall mad with my trilobite obsession. He ran away to the States as soon as he could, while I continued to learn the messages of the trilobite shells.

Most of the information about trilobites was buried. The first task I had to complete was to excavate the specimens from their rocky matrix, and I spent months doing just this. My original "lucky break" at St. David's was unusual because the trilobite split out more or less complete: this does not happen often. Usually you might see the top of the glabella, or maybe an eye; you just have to grind away the rock that surrounds the animal to excavate the hidden truth, and that is done later in the warmth of the laboratory. It is a skilful business, and you acquire the skill at the expense of heartbreak. Little mechanical percussion needles are a standard tool—they make an insistent buzzing noise like that of an enraged wasp—but one slip and you gouge a terrible wound across the face of the trilobite you've grown attached to. You rely on the fact that the rock has natural tendency to split around—rather than across—a fossil. Sometimes you get it wrong and a lump of your precious find flies across the room, leaving you grovelling on the floor with a magnifying glass looking for the detached fragment. Some days I would spend hours with a dissecting needle peering through a microscope and flicking off minuscule bits of matrix to reveal the beast beneath. John Bursnall used to accuse me of carving out fossils to my own design.

The best needles were those used to play the old 78 rpm gramophone records, which were already a rare commodity in the early 1970s. My fellow student Phil Lane and I used to hunt junk shops for these sharpenable, but hard, steel needles. If we found a cache we could purchase it for a few pence, to the mystification of the shopkeeper. "Can I interest you in some old records to go with the needles?" he would say. "No thanks, just the needles," we replied, heading speedily for the door and trying not to look as though we intended to use them for experiments with drugs.

It was soon evident that most of my trilobites were only parts of the animal. Whole ones like my first lucky find are actually rather rare. The carapace of the animal often fell to pieces after it died, like a suit of armour breaking along its joints. Whole animals do not hang together for long. The tail-shield is rather robust—an isolated pygidium is often the first thing you find when you split open a piece of rock. The most fugitive part is the thorax, which disarticulates into its separate segments, which are then broken or dispersed. The cephalic shield also frequently breaks into several pieces. A middle part carrying the glabella makes up one of these fragments. It is called the *cranidium*. Either side of the cranidium is flanked by a *free cheek*, or librigena—left and right, and mirror images of one another. Many trilobites carry a prominent spine at the edge of each free cheek which form spiky hind corners to the head—these are *genal spines*. The eye surface is attached to the free cheek in most trilobites: indeed, the cephalon is *designed* to split into three or more pieces—including cheeks and cranidium. The free cheeks are separated from the cranidium by special planes of weakness—suture lines. These facial sutures were an aid to the animal during moulting. The eye surface was probably the most vulnerable part of the trilobite, and difficult to moult successfully. By splitting the headshield along sutures which run fore and aft, and include the eye surface at about mid-length, it was possible to shed the eye surface before moulting the rest of the carapace. This speeded up the whole business and reduced the time the animal spent in the "soft shell" state. The cheeks were liberated first, and independently from the rest of the animal, and hence were "free." The *fixed* cheeks remained behind, next to the glabella on the cranidium.

So the average trilobite yielded a considerable number of bits and pieces as they fell apart: cheeks, thoracic segments, cranidium, pygidium. And because they moulted several times as they grew during their lifetime, casting their old shells and growing new ones just as crabs and lobsters do today, it is obvious that a large trilobite would have to shed a succession of its earlier coats as it gradually reached its mature size. All of those earlier pieces were potential fossils—trilobites were veritable fossil factories.

But this is where the problems come in. If you only have the pieces then your first job is to reconstruct the trilobite as it was in life. It is like having to do a jigsaw puzzle of uncertain design. Worse still, if you have a dozen or more species of trilobite jumbled together and all of them fragmentary it becomes more like doing a dozen jigsaw puzzles simultaneously without having a colour picture on the cover of the box. While I was learning my trade, I became adept at matching little pieces. There were clues. The shape of a suture edge on a free cheek should match that on the corresponding cranidium. Then my predecessors had described occasional whole animals in their publications, and once I had recognized the cephalon, say, as belonging to a particular type of trilobite I could consult these illustrations for a search image of the appropriate matching pygidium. Quite soon my office was a jumble of broken bits of rocks, and needles, and old monographs, all coated in fine, limy dust. I still work in an identical office today. Tidy people's eyes go all peculiar when they come into it. I have a special small padded seat for them to collapse into.

It was fascinating work, more like that of an archaeologist gluing together potsherds than anything to do with white-coated science. From time to time Harry Whittington would appear and make encouraging remarks, or put me right when I had placed the wrong head and tail together. He was the gentlest of supervisors, surely better described by the American term the PhD "advisor." His advice was always welcome. He had written the papers and monographs that I most frequently thumbed, and nowadays my personal copies have lost their covers and become dog-eared through years of use.

Harry Whittington has probably added more to our detailed knowledge of trilobites than anyone else. In the 1950s he discovered some remarkably preserved shells. In the Edinburg Limestone, an Ordovician limestone which can be found in a few roadside exposures in Virginia, the trilobite shells have been replaced in almost perfect detail by the hard mineral silica. Since the matrix of the rock is made of limestone it can be rendered into solution by throwing blocks into dilute hydrochloric acid. The limestone is often rather dark in colour, and when you place the piece into the acid it fizzes violently, like an Alka-Seltzer. Gradually it settles down to a regular bubbling—more like soda water. Then you notice that little ridges are starting to project from the lumps of rock—these morsels are not dissolving. They are silica trilobites being etched out of the rock. When the process is complete the fine mud and muck can be washed away through a sieve, and what remains is pure trilobite.

This is like suddenly having the living shells at your disposal—handfuls of them. It is as if you have been transported back in time through more than 400 million years and given the freedom of an Ordovician beach. You can turn the pieces of trilobite upside down and look at the *under*side of the shell for the first time. You can inspect the doublure. What might take weeks to prepare manually from the rocky matrix, can be etched out incomparably well in a matter of days. The larger pieces can be picked up with fine tweezers, the tiny ones with a moist paintbrush and transferred to slides. What you discover is a pile of different free cheeks, pygidia, cranidia, and thoracic segments, a matted palaeontological bonanza. Then they all have to be picked apart under the microscope and matched together, as one might sort with excited anticipation through a spectacular box of bric-a-brac picked up at a jumble sale.

What Harry Whittington's preparations revealed was the fantastical sculpture of some of these trilobites. Spines and prickles—indeed, prickles on top of other prickles—were all picked out perfectly by the silica replication. Even the most careful preparator could never do justice to this thorny magnificence. Some of these trilobites were hairier with spines than any hedgehog. They covered the head, sprouted off every thoracic segment—often splaying off the edge like a defensive comb of slender rapiers—and continued on to the pygidium, where one pair might continue backwards beyond the rest of the animal and sprout subsidiary protuberances. Surely these were animals as idiosyncratic as any seahorse or spider crab? Even more remarkable was the preservation of

Amazing spines recovered from the perfectly silicified (see p. 38) trilobites from the Ordovician of Virginia. This is a cranidium (top right), pygidium (bottom) and free cheek (top left) of a trilobite related to *Odontopleura* called *Apianurus*. Even some of the spines have spines! Such minute details are almost impossible to excavate manually. (Courtesy H. B. Whittington.)

diminutive organs on the surface of the animals. It was possible to see that the tips of many of the spines carried tiny perforations. In life, maybe, even tinier sensory hairs emerged from the minute holes, sensitive monitors of their ancient marine world, quivering to the scents and vibrations of the time.

Other trilobites had a surface covered with ridges as complex as a fingerprint, swirling in concentric arcs, exuberant as a Jackson Pollock painting. Others again were covered in tubercles, little round lumps that made the surface of the shells look as if dusted with dewdrops. Then there were those that instead of having any projections had the surface of the shell dotted with tiny pits. One trilobite we shall learn more about was surrounded by a kind of perforated sled forming a fringe around the whole of the cephalon. All these were identified and pieced together by Harry Whittington when he sluiced out his acid vats, washing away the millions of years of entombment from the shells as he poured away the Ordovician muds that had once buried them.

There are some other shelly pieces left behind on the sieve that did not fit into the now familiar inventory of head, thorax or pygidium. Most noticeable are a variety of oval plates surrounded by rims, usually with a pair of long projections at one end. It is known from the study of complete trilobites that this plate fitted on the underside of the animal in the middle of the cephalic shield. The plate was a backwards continuation of the marginal undershelf—the doublure—at the edge of the trilobite skeleton. It is called the *hypostome*. It corresponds quite closely with the front part of the glabella—but lay on the bottom rather than the top of the animal. Whatever was within the glabella was accordingly well protected by a calcite skeleton above and below: it must have been something important to the animal. In fact, this vulnerable region included the brain, such as it was, and the stomach—together, the two most vital of the vital organs.

All these shelly pieces of the trilobites were part of what is called an exoskeleton, because they lay *outside* the soft anatomy of the animal. They were the crisp wrapping of a succulent parcel. Just the opposite is true of *Homo sapiens* and his vertebrate relatives, in which the flesh is hung on the bones—the soft parts on the outside. Man is designed so that he can be stabbed in the back. Trilobites and other arthropods purchased their invulnerability to treachery at the price of having to change exoskeletons with growth, shedding every last tiny spine and tubercle and growing them all over again in a new suit of calcitic clothes. The hypostome was shed at the same time as the rest of the shelly paraphernalia.

Like the silica trilobites he studied, Harry Whittington himself has defied the passage of time. As others fade, he carries on indefatigably studying his beloved fossils. I like to think that this is a reflection of his own virtues, the unfading ones of kindness and perseverance. His hair and moustache are still hardly tinged with grey and he is 83 years old. He is a native of Birmingham, in the middle of England, but spent many years at Harvard University, which has left him with an indefinable accent—not exactly transatlantic, but unplaceable. At the time he became my mentor he had returned from the United States to become Woodwardian Professor of Geology in the University of Cambridge—one of those ringing titles which are so characteristic of ancient seats of learning. A polished label announced this distinction on the door of the old office that had been occupied by Woodwardian Professors for more than a century. It could have been fusty, but I always felt that the place constituted a link with previous generations of scholars, and thence back through uncountable ages to the time of the trilobites. Somehow one would not have been very surprised to have been greeted there by the ghost of the Revd. Adam Sedgwick, the nineteenth-century Cambridge geologist who coined the term Cambrian—the strata from which my very first trilobite had been wrested.

Harry Whittington often went on fieldwork with his wife, Dorothy. She was as ebullient as he was quiet, and exemplified a remarkable principle of discovery: partners always find the best specimens. Whittington and his students would be squatting in the floor of a quarry bashing at some grey, tough limestones with their geological hammers. Thumbs would have been struck, oaths uttered. From time to time a tantalizing fragment would be found, sufficient to keep everyone beating the bedrock in a frenzy of curiosity. Dorothy in the meantime would pick at the odd piece of strata in a leisurely fashion, all the while enjoying the spring sunshine. Then would come the question: "Harry, is this anything?" Sitting in the palm of her hand would be the gem of the day.

Harry Whittington is an authority on trilobites. There is a great difference between authority and authoritarianism. Some professors qualify under both headings, but the best are those whose authority is earned through the regard of their fellows: *primus inter pares*. I have met some of the other variety, too. When I was a visitor at the University of Göttingen in Germany I went to the coffee room at the usual time in search of refreshment. Seeing a spare chair at the table I sat down upon it and started sipping my coffee. A terrible hush fell upon the room. It seemed to have something to do with me. Puzzled, I checked my fly, and looked for other signs of disgrace. The wooden chair I sat in seemed to be identical to a dozen others around the table. After a minute of excruciating embarrassment, one of the younger men in the Department came over and whispered in my ear: "That is Herr Doctor Professor *****'s Chair!" *Mein Gott!* I leaped to my feet, blushing red from my collar upwards, and quickly found myself another, apparently similar chair. That is authoritarian.

I returned to the island of Spitsbergen in 1972. The trilobite finds on which I had worked had proved so exciting that the Norwegian government was prepared to finance a full-scale expedition to the far northern part of the island to collect still more, and to fill in the gaps that had been left unexplored. This was a grand affair compared with my earlier visit, when there had been just the two of us, plus a tent, a small boat, and as much porridge as we could eat. The remote shore looked every bit as bleak on my return, an endless stretch of shingle swept by an uncompromising wind. I recognized the melt stream which ran off the great glacier occupying the middle of this part of the island. We had camped by it before. Arctic terns shrieked a neurotic welcome. Now we had a team of eight or so, and a rather grand tent, almost a marquee, for communal evenings and sitting out the blizzards. It could be heated up to a delightful cosiness. Flitches of ham hung from the roof, alongside intriguing salamis. There was a sophisticated radio system and another kind of ham to operate it. We could sit at a trestle table in the evening and exchange the sort of banter that keeps an expedition on its toes. Once in a while tempers would fray. I concentrated hard on being everyone's friend.

We had another kindly professor with us—Gunnar Henningsmoen, from Oslo—possibly the only rival to Whittington in his generosity of spirit. He presided over our suppers with unfailing good humour. I shared my little tent with David Bruton, the only other Englishman, who had long since taken up residence in Norway, and who spoke Norwegian with gusto to everyone else. For reasons of outmoded chauvinism the two of us insisted on flying a Union Jack outside our tent, which then slowly unwove and unpeeled itself over the next few weeks until it was the merest tatter: so much for the British presence. The oddest experience for a foreigner like myself was not being able to share a joke around the dinner table—for jokes do not translate, and anyway they are born of the moment, and flag with repetition. You sit there with a weak smile on your lips hoping to demonstrate a sense of humour even if you have no idea what everyone else is laughing at. (You hope it is not a joke about *you*, but you would sit there with the same silly grin even if it were.) Little bits of Norwegian came to me by a kind of aural osmosis. The most surprising linguistic fact I learned was the impoverishment of that language in swear words. In fact, there *is* only one—*"farn"*—which merely means something like "devil take it!," but is considered very rude by a well brought-up Viking. It has to pass muster for most of the everyday tragedies that beset an expedition. If a finger is hammered, you jump up and down and cry "farn"; if you drop an outstanding fossil irretrievably into the sea, you splutter for a while and then mutter "farn" under your breath. If all your provisions were carried away by a hurricane and death were guaranteed, all the poor Norwegian could do would be to stand on the shingle and cry "farn" into the wind. Somehow this does not seem adequate for the occasion.

We collected box after box of specimens. Sooner or later they would find themselves the subject of scrutiny under my binocular microscope back home. My small personal fragment of history—ten million years or so—was that distant time when the rocks enclosing my trilobites had been laid down. I could wander around that time in my mind as comfortably as a historian might amble among the Tudors and Stuarts. I could match the various trilobite pieces together more quickly than anyone else: cranidium with free cheeks, pygidium with cranidium. Once in a while somebody would find a complete specimen, which was like suddenly finding the lid of the jigsaw puzzle. It was a way of testing your own earlier inferences about what fitted with what. I discovered an extraordinary, goggle-eyed trilobite that I called *Opipeuter inconnivus*—which means "one who gazes without sleeping." It could have been a description of me. Slowly, a picture of a vanished Ordovician ocean was forming in my mind. Why, there were probably many more species then on the site of my bleak shore than were able to live there today! The Ordovician sea was a rich one: just because it was ancient did not imply that it was impoverished. At that time there was virtually no life on land, but the sea thronged with jellyfish and trilobites, clams and snails and segmented worms. There were fierce predators related to the living pearly *Nautilus*. There were groves of seaweeds. There were even shoals of small, lithe animals that might, at first glance, be mistaken for silvered masses of fish. A palaeontologist does not only listen to the message of this fossil animal or that. What he does is to recreate a vanished world.

I was invited to give a lecture on the glories of the new Spitsbergen discoveries to the Norwegian Academy of Sciences, as august a body as you could wish for. Norway has special sovereignty in this part of the Arctic, so my invitation was not untinged with politics. It was an intimidating experience to stand before an audience of a hundred or more of the most distinguished scientists in Norway, a sea of sages. To be twenty-five years old and on stage in a fine and historic building in Oslo is not such an easy occasion to mark the transition from taught to teacher. The great Arctic explorers Nansen and Amundsen had stood on that same stage, and portraits of other notables looked on. It was just as well that there was so much to talk about: the surprise discovery of some of the richest fossil faunas in the world on the remote shore of Hinlopen Strait; why others might have missed the rocks before; how the trilobites proved a connection of Spitsbergen with the ancient continent of Laurentia; how the climate at Ordovician time had been tropical rather than Arctic. This was my first public conjuring of the excitement of deep prehistory. Once the adrenalin started to course through my system, the audience became reduced to a hundred sets of ears.

At the end a tall and courteous old man stood up and asked a question in impeccable English. He referred to his time on Novaya Zemlya in the early decades of the twentieth century. His name, he explained, was Olaf Holtedahl. I was astonished. It would scarcely have been more amazing if Fridtjof Nansen himself had stood up and asked me about my Arctic experiences. Holtedahl was a survivor from the heroic generation when Arctic exploration had been truly a trip into the unknown, a time when huskies were the main means of transport, and pemmican the main source of protein. In the twenties, he had written pioneering reports of the geology of the high Arctic, and particularly concerning the remote island of Novaya Zemlya, which points northwards into the Arctic ocean from the Russian coast like a crooked finger. Not much had been published about this island since his pioneering visits there—and what there was was in Russian, because it was a highly secret military area during the Cold War. So here was a figure of romance and scientific derring-do who had stepped off the page, from the realm of my imagination, and into the flesh, dressed in a rather immaculate suit.

This incident made me understand a connection with a past different from that remote one reached by listening to the messages of the shells: the past of my scientific forebears. In the egotism of investigation there is a tendency to forget that there were students before us who made the discoveries on which our interpretations still depend. Science is an odd venture, at the same time co-operative and competitive. The motive force is often the desire to beat a rival investigator to the credit for a discovery. But in the long term such human rivalries recede, and what started as a race seems more closely to resemble a series of logical advances linked together by a roster of names of discoverers.

The first name on the trilobite list is exactly 300 years old as I write: Dr. Lhwyd, whose letter to Martin Lister concerning "the sceleton of some Flat-Fish" opened this chapter. The letter was published in 1679 in the *Philosophical Transactions of the Royal Society*, the oldest scientific journal in the English language; the title of the article: "Concerning some regularly figured stones lately found, and observations of ancient languages." I like to think of these misplaced "flatfish" rubbing shoulders in the same journal with the reports of the pioneer microscopist van Leeuwenhoek listing the discovery of red blood corpuscles, and microbes, and other such momentous stuff. Trilobites were smuggled in as observers of mighty things from the very first. The early volumes of the *Transac-

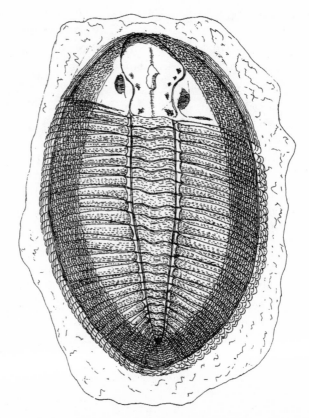

The "flatfish" figured in the *Philosophical Transactions of the Royal Society* by Dr. Lhwyd (1679)—in fact, the trilobyte *Ogygiocarella debuchii* from the Llandeilo (Ordovician) rocks of South Wales. The trilobite itself is photographed at fig. 1.

tions are treated with some reverence, as they should be; the best leather bindings are no more than they deserve.

Those who know the rocks in the vicinity of Llandeilo are in no doubt as to the identity of the "flatfish"—it is a trilobite called *Ogygiocarella debuchii* (see above). In many places around the castle—Dynefor Park, just outside Llandeilo— there are pits where masses of flat-lying, limy flags are exposed; they can be pulled from the hedge banks in slabs like so many polygonal plates, and on some of these plates flatfish are indeed served up: the size of small plaice, and nearly as flat, they have two eyes goggling up at the surprised collector. The modern observer can perceive that they have a thorax with eight segments and a large pygidium, too, and therefore that they are no manner of fish—but you can see how Dr. Lhwyd made his mistake. He embellished the drawing a little to show something resembling a marginal fin. He was right only about the eyes.

Trilobites were recognized as a distinct group of animals in 1771 by a German zoologist called Walch, in a work of such obscurity that I am still not sure whether we have discovered the right edition in any British library. But within ten years articles by such scholars as M. T. Brünnich were using "trilobite" on the title page, so it must have achieved wide currency—and it is, after all, a euphonious and descriptive name. More and more of these distinctive "organic remains" had evidently been discovered across Europe. By the first two decades of the nineteenth century many more trilobites were receiving scientific names, especially in Scandinavia, France, and Germany. Lhwyd's animal attained its appropriate recognition in 1822 when the French palaeontologist Alexandre Brongniart published a short treatise on our animals, *Les Trilobites*, in which he named the species *debuchii* from a specimen from Lord Dynefor's estate. The flatfish had finally vanished, and in its place was a strange animal with a calcareous skin and segments like a lobster.

One hundred and forty years after the note in the *Philosophical Transactions* Lhwyd's "flatfish" would be used to recognize and correlate rocks all the way between Llandeilo and Shropshire. In Sir Roderick Murchison's book *The Silurian System* (1839) trilobites like *Ogygiocarella debuchii* are illustrated not only for their interest, but also for their utility in identifying rocks of a particular age. By then the name "trilobite" had achieved a certain familiarity among the educated classes which it would never lose again. The classical names which guaranteed their identity as animals were given when all scholars knew the *Aeneid*, and were as familiar with mythology as we might be with the cast of *EastEnders*. *Ogygiocarella* was named for Ogygia, the seventh daughter of Amphion and Niobe. And in turn both of these names from Greek mythology were attached to other trilobites. It is actually quite difficult to find a classical name that has *not* been used for one animal or another, be it an ever-so-obscure Phrygian nymph or a goatherd from the flanks of Mount Olympus. There are layers of time and levels of antiquity: there is a primal layer, the original ancient time of the trilobites; then there is the Greek or Latin source of the names—the time of the "ancients"; then there is a history of research; and finally there is a personal history which animates all those earlier times in the shape of the specimen to hand.

It was not long before similarities between trilobites and living animals were noted. Segmented creatures were among the fauna that could be found crawling around on seashores or forest floors—indeed, they were among the commonest of animals. Insects, crustaceans, spiders, centipedes—all were constructed from linked segments which articulated with one another. They shared another common feature: jointed legs. At first glance, it may be difficult to appreciate the similarities between the legs of a fly and the legs of a lobster. But they are jointed in a similar fashion so that each joint can rotate or swivel relative to its neighbours in a programmed way according to exactly how they are hinged. They have the slightly infuriating predictability of one of those jointed reading lamps: you soon discover that they will swivel in a finite number of ways, but nonetheless it is possible to direct them into the most improbable corners once you have mastered the joints. You can get the measure of the range of possible movements if you hold a struggling lobster upside down—the legs kick in and out in a mechanical way. Watch a beetle kicking its legs when stranded on its back and the similarity is evident.

The meat in these animals is all inside the leg; muscles contract to effect movements along joints. They pull themselves up by their own internal bootstraps. Jointed-legged animals are called arthropods, and there is no doubt at all that trilobites were another kind of arthropod (even though it was a long time before their legs were found fossilized); had they survived they would have been lined up alongside scorpions and crabs, butterflies, beetles and bedbugs as another example of the most diverse and varied of all animal designs. Carl von Linné (or Linnaeus), the father of biological classification, had recognized the trilobite pedigree even before the close of the eighteenth century. Had they not died out, I imagine that on the beach mothers would plead with their children: "Jimmy, please don't pull the legs off that poor trilobite!" Jimmy would not have been able to resist the temptation to wiggle the limbs of his captive beast to see how they could be bent preferentially in certain directions. He would wave the creepy-crawly trilobite about to scare Aunt Margery, who cannot stand that kind of thing.

But the trilobite shells I studied in Spitsbergen were only empty carapaces. Lacking a calcareous coating, the limbs had vanished into oblivion. I could almost feel the tickle of legs crawling over my palm, I could imagine the living animal scuttling through the Ordovician seas. In my mind's eye I could cross-breed the trilobite with a prawn or a scorpion according to my fancy. But there are places—rare places— where such speculations can be fleshed out with real specimens, where even the most delicate and fragile hairs on the most spindly of appendages are miraculously preserved. It is to these places we must go if we wish to discover the whole truth about trilobites, and allow them to tell the tales they have to tell.

III

Legs

If you wish to snare a rare butterfly it is no use using a blunderbuss and a suitcase. To search for something elusive, subtlety and intelligence is required—and no small measure of luck. Most quests for unusual goals are motivated by the conviction that the end is attainable if only the correct combination of fortune and persistence can be struck. So it was with the case of the trilobite legs.

By the middle years of the nineteenth century several hundred different trilobites had been named and described. These were heroic years for monographs. Geologists were mapping and exploring ancient rocks in a systematic way for the first time. They were beginning to understand geological time, and creating names for its divisions, many of which we still use today. As they mapped the strata in succession they recognized the utility of fossils in discriminating strata of a particular age. The sequence of fossil species was a thoroughly pragmatical way of unscrambling the chaos of the strata. In Britain, the explorers of trilobite-bearing strata were tramping on foot or progressing in carriages across Wales; they were true pioneers, the first to crack open ungrateful shales. In North Wales the Revd. Adam Sedgwick—in whose eponymous museum I studied in Cambridge—named the Cambrian

System (Cambria—Roman name for Wales) for ancient rocks that underlay all the other fossil-bearing strata—the most fundamental package of geological time. The name Cambrian eventually replaced a rival name, "Primordial," which hinted at the fundamental position of the organic remains these strata contained in relation to the subsequent history of life. While Sedgwick explored North Wales, Sir Roderick Murchison was crossing South Wales, mapping and classifying his Silurian System (1839) (Silures—a tribe that once inhabited South Wales). Even more than Sedgwick, he used the fossils to spell out the narrative of geological time. Trilobites were easy to recognize; they became the familiar face of various parts of the Silurian. It is easy to imagine Sir Roderick imperiously summoning the rock-hunting vicar of some distant Welsh parish to present his discoveries from his local stream or *cwm*—and a smile playing on the aristocratic lips as he identified *Trinucleus fimbriatus,* or some other old trilobitic friend. Just as a numismatist might intimately know a seemingly freshly-minted coin of Hadrian, so the palaeontologist would greet the familiar face of a species first encountered fifty miles away and a decade before. Geological time was written in a thousand trilobites.

The palaeontologists Henry Hicks and John Salter were the first to tread those cliffs that I visited as a schoolboy in Pembrokeshire. I have a photograph of them in the 1870s grinning in comradely satisfaction at their latest discoveries. The trilobites they found still bear the imprimatur of their names. When we cite the Latin name of a species we are obliged to add the name of the scientist who first gave it its official name. *Ampyx salteri* Hicks was an interesting species from the dark slates exposed in the cliffs north of St. David's, first described in a publication of the Geological Society of London in 1873. Here Henry Hicks presented a gift to his friend—he named a trilobite species for him: Salter's *Ampyx.* Salter reciprocated by naming one of the beautiful Cambrian trilobites from the Pembrokeshire coast *Paradoxides hicksi* Salter. I have done it

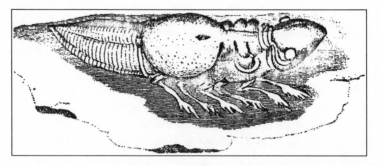

A fantasy trilobite from J. S. Schroeter (1774)—the first portrayal of legs. This animal combines one head (right way round), one head (reversed) and a tail (possibly upside down) with purely speculative legs.

myself: to celebrate the devotion of Frank Cross to collecting trilobites in Welsh ditches in all weathers on behalf of myself and my friend Bob Owens there is a delightful small species called *Shumardia crossi* Fortey & Owens. Our debt to Mr. Cross will thus be recorded in perpetuity.

Wales was only one area which was the subject of intense palaeontological scrutiny in the period 1830–75. James Hall was using every means, fair and sometimes foul, to secure publication of his great monographs on the Palaeontology of the State of New York; and in Bohemia Joachim Barrande was calibrating the same time interval with different fossils. But as first dozens and then hundreds of different trilobites were discovered, and the bounteous variety of their form became apparent, so too did the great gap in knowledge of the animal itself. All the fossils were shells, husks, mute carcases that could not speak fully of their vanished lives. There was no trace of the jointed limbs that every observer felt should have propelled the animals along the sea-bed. Some early investigators simply made them up (see above). Without their limbs these wonderful fossils were geological cyphers, patterned stones that spoke only of geological time, as passive in their way as that coin of Hadrian—no more than a token of real life.

Legs

Until their legs were discovered it was impossible really to know the trilobite.

There had to be a way to discover the identity of trilobite limbs—but how? Their covering must have been the same thin sheath of the organic polymer chitin, which covers the legs of shrimps and centipedes today. This does not fossilize as readily as mineral shell does, but it is not impossibly evanescent, like the glutinous shadow of an amoeba, say. There must be circumstances that might favour the early entombment of these limbs in a sediment soft or special enough to favour the preservation even of a wisp, a shadow.

There were hints. Many species of trilobites were capable of rolling up tight into a ball (see fig. 16). This is a method of protection in dozens of living animals—even humans instinctively curl up under a hail of blows. The hedgehog protects its vulnerable belly this way, and thereby renders itself pathetically liable to be crushed under wheels for which evolution had never prepared it. The best comparison of all is with many species of little isopod—crustaceans that swarm under rotten wood. Almost any old woodpile will yield them in their hordes: turn over a rotten log, and their miniature armoured capsules will crawl rapidly away from the light. They are called woodlice, slaters, or pillbugs according to where you come from. Like trilobites they have a shell on their backs and vulnerable legs. Many of them adopt the same protective strategy—they roll up. I have seen some woodlice so tight and perfect that they resemble ball bearings—they even have a sort of sheen. Their segments are perfectly designed to slide past one another in enrolment. Their legs are adapted to tuck up inside, like stowing oars inside a boat. Although not closely related to trilobites (other than being fellow arthropods) they offer a useful comparison. Many trilobites are equally tightly enroled—but bigger. You can hold a rolled-up *Symphysurus* (see p. 188) in your hand like an egg: it has the same satisfying feel to it. Look closely, and you can see how the edges of the thoracic segments have telescoped together, the tips sliding past one another like the blades of a Japanese fan. The thoracic segments have special facets to permit this. At the same time the axis of the thorax has extended, and here an extra "half-ring" is revealed between the segments to cover the gap that otherwise would open up. You can see a similar arrangement on the flexible elbows of a suit of armour. Clearly these animals took their enrolment as seriously as a jousting knight took his protection from sneaky blows. Some trilobites even had little locks to make sure that the fit was extraordinarily tight.

Inside one of these enroled carapaces is it not possible that the stowed limbs would be preserved? This could be the true time capsule. What was needed was a specimen killed even as it enroled, and then preserved in sediment quickly— maybe beneath an ash fall akin to that which buried the unfortunate inhabitants of Pompeii and Herculaneum. Such volcanic eruptions frequently showered into the sea in ancient times, causing mass mortalities. It would also be important that the rolled-up ball should not have been distorted or flattened after burial in the strata. It seems a lot to ask, but well preserved, whole, encapsulated animals are known from a number of localities—they were recognized long ago in the Ordovician limestone rocks of Sweden and Estonia, and the Silurian ones of England. The trilobites were sliced through and then polished patiently with emery powder. Surely the traces of those elusive limbs would show up on the polished section. Sadly, it was not to prove so. The enroled balls were filled with fine sediment, which must have oozed in after the animal was buried but too late to save the limbs from obliteration: the bacteria got to work and left nothing behind. Quite possibly the bacteria lived in the very sediments that came to occupy the capsule. Maybe, just maybe, on one or two specimens, a hint of a leg cross-section marked by a delicate dark circle, a suggestion of something else . . . but the mystery of the detailed structure of the delicate limbs remained unsolved.

In 1876 a young palaeontologist called Charles Doolittle Walcott (1850–1927) made the first progress in solving the riddle. He was an avid collector of trilobites in the vicinity of Trenton Falls, New York, and one of the most remarkable self-made men in a century when self-improvement was a watchword. He came from a farming family, stolid and god-fearing and with no particular intellectual bent. Without any formal geological degree, he progressed to become the Director of the United States Geological Survey and Secretary of the Smithsonian Institution in Washington, DC. He undoubtedly had the most inappropriate middle name in biography. For as well as becoming the ultimate Washington operator, cultivator of politicians and professors alike, and a very busy administrator, he found time to produce a great series of publications on trilobites—and many other kinds of fossils besides. His books fill a library shelf. He named dozens of the trilobites which calibrated the Cambrian strata of the whole North American continent. He investigated the succession of rock formations in the Grand Canyon, often under the most punitive conditions. This was before the trails had been cut that make the geologically aged depths accessible today—and even now you see exhausted walkers who have collapsed by the track because they have ignored all the notices about bringing water with them. After all, who really believes that wilderness can be only an hour or two away from the luxury of a Holiday Inn?

Walcott is most widely remembered as the original discoverer of the celebrated Cambrian Burgess Shale, in British Columbia, but, even if he had never made this find, his place in the history of science would have been assured. The modern worker gapes in wonderment at Walcott's output. "Well, he did not have to answer the telephone all the time," his contemporary counterpart might grudgingly allow, "and in those days he could afford to live near the office in DC." There follows a diatribe on the price of real estate. All true, no doubt, but it does also seem to be true that a century ago there were more people like Walcott, with an extraordinary capacity for hard work which transformed talent into achievement by force of will. One thinks of Sir Walter Scott grinding out his great pile of novels (and that partly to pay off his publisher John Murray's debt). Method was no doubt part of it; I am sure that Walcott always knew where he had placed that important piece of paper on the previous afternoon. He side-stepped those diversions that always persuade us to delay until tomorrow what we have already deferred from the day before yesterday. He even found time to compile a diary on a regular basis—another activity which is most often the subject of good intentions rather than consistent practice. I only wish that the diary entries were more interesting—for the most part they reveal nothing of the man, but a lot about his appoint-ments with the influential. He exposes raw emotion only after the tragic early death of his first wife, Lura, at a time when he was yet to receive his first professional appointment. As he rose to greatness the entries became more and more perfunc-tory. But it was in the weeks following his bereavement, when he worked incessantly to keep grief at bay, that he made the discovery of the elusive trilobite limbs.

Around the town of Trenton Falls there are limestones which crop out sporadically along road cuttings or in the sides of streams. These are the strata that Walcott knew so well. The seams have also been exposed in lime quarries and other pits, which these days are often in danger of being infilled with the profligate detritus of the consumer society: a hole is often more valuable than what was once extracted from it. The Ordovician limestones are exposed in bench-like beds, and in contrast to the contorted shales of Wales or Cornwall, these strata are usually horizontal, or nearly so. It is clear that such rocks have not been mangled by convulsions of the planet such as those that elevated ancient mountain chains. They record, virtually undisturbed, a succession of ancient sea-floors, one after another, which the young Walcott was lucky enough to explore for the first time. In some cases a single rock

Charles Doolittle Walcott (centre) in the field.

bed an inch or so thick was the product of a storm that culled
the life of a vanished afternoon—and afterwards the community of animals would re-establish itself by the time of the
deposition of the next rock layer. Even after more than a century of collecting, the abundance of fossil shells is still exceptional, and in Walcott's day it must have been astounding.
Often a piece of rock resembles a slab of fruit-cake with trilobites and brachiopods and snails and sea mats, and many
more organisms besides, dotted over the surface like so many
plums and raisins and sultanas. You want to pluck them out,
but they are firmly lodged in the surface, and dusted with a
paler limestone which may obscure their finer details. Patient
cleaning with a pin will clean out the most magnificent examples. Charles Doolittle Walcott could have first pick.

On 1 March 1876, just over a month after Lura died, Walcott wrote in his diary: "Cut up several C. ps. had fair success.
I think I shall determine their interior structure." This was
nothing less than his note that evidence of limbs might be

preserved in "C. ps." The abbreviation stands for *Ceraurus pleurexanthemus* (you can understand why he might want a shorthand!), a trilobite with a spiny pygidium and a knobbly glabella, which is one of the most exciting finds that can be made from the Trenton rocks. It had been known as a carapace for more than forty years, since J. Green had named it as one of his in *A monograph of the Trilobites of North America* in 1832.*

But there was no hint that this species would offer the key to fleshing out the reality of trilobite limbs, until Walcott sectioned and polished the "Ceraurus layer"—a two-inch-thick band of limestone in which these animals were common. This single rock bed had many individuals of the same species. At the bottom and top of the layer were beautiful, entire, but empty individuals—shells alone—the ornament of many a collector's cabinet, but no more informative than a thousand other specimens. The animals seem to have been trapped by an sudden inrush of limy sediment such as might have followed upon a tempest—a minor tragedy in the life of a generation of trilobites, but a boon to the scientist who followed 440 million years later. In the middle of the layer were animals that were overwhelmed and killed; some died struggling towards the surface of their sediment tomb. One can imagine a quiet sea-floor on which trilobites crawled about in profusion; suddenly the water darkened with a blanket of fine mud that settled in a choking blanket before some of the animals could effect an escape. The poor things had no time to roll up completely, but many of them had started to flex their bodies, curving towards an enrolment they never achieved at the moment of suffocation. Perhaps the idea of the enroled time capsule was half right, after all. The limbs of these animals trapped in mid-layer did not immediately rot as did those of

---

*This is a most peculiar book, for it came with a set of coloured models of the species described, which can still be found in the drawers of some of our oldest academic institutions. Green hoped that the sale of the sets would make an additional attraction for the purchaser. Sadly, some of the reconstructions are somewhat approximate.

animals near the top. Instead, there was time for them to become filled with lime-charged water. What had once been filled with muscle became replaced with white calcite. Like the mummification process perfected by Pharaonic priests, a preserving ichor flooded the soft appendages. The mineral calcite filled the limbs so that, even when the tissue surrounding was eaten away, infilled moulds remained as testimony of the soft parts. Walcott had the insight to recognize that when his polished sections revealed tiny white circles which contrasted with the surrounding grey limestone, these were the infilling of the arcane appendages. Within a few days he had repeated his observations on another species. "Found that *Calymene senaria* has the same character of appendages that C. p. has. Wrote description of C. p. after supper," he recorded in his laconic diary entry on 10 March 1876. One can imagine that he would have reported almost any discovery in like fashion: "Found Holy Grail this a.m.; have expectations of Excalibur tomorrow." Trilobites were never the same again.

Consider what Walcott still had to do. The discovery required him to grind—by hand—a whole series of cross-sections. He needed to assess how many limbs there were, and whether they branched or were simple, jointed legs. Limestone is not soft (tap your nearest Victorian fire surround to prove the point), and he used a cutting wire and a rotating lap to cut and polish his sections—a slow business. And, most difficult of all, he needed to match drawings of sections one to another to obtain a three-dimensional picture, something that in modern laboratories even computers find challenging. One can imagine him assuaging his grief by working on late into the night, curiosity and ambition displacing other, darker thoughts. Nonetheless he managed to write his report of his discoveries, which was published in preliminary form before the year was out. It carried the ponderous title "Preliminary notice of the discovery of the remains of the natatory and branchial appendages of trilobites." While true to the letter of the discovery, it could scarcely be a best-seller with that title.

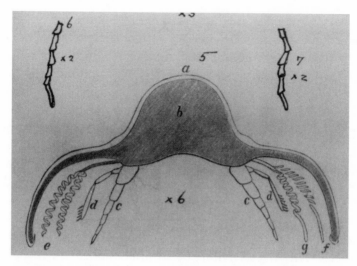

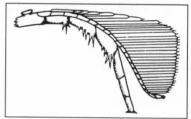

Charles Doolittle Walcott's first, and inaccurate, attempt at portraying the limbs of trilobites *Ceraurus* discovered from sections ground through rock surfaces. Right, a modern reconstruction of the branched limb of *Triarthrus* (after H. B. Whittington and J. Almond).

In any case Walcott continued to make new observations on further sections cut over the next eighteen months, and modified his original reconstruction. But the first version shows several interesting features. Walcott had noticed one important fact that still remains true. There were paired appendages on each segment, and so far as he could judge they were rather similar along the length of the animal. The appendages hung down below the body cavity, where the "guts" were. This visceral mass was largely under the axis of the animal—

the middle lobe of the "tri-lobes." So the legs and other appendages operated underneath the body of the animal, safely enclosed by the sloping parts of the pleural areas. Towards the middle were jointed legs—the kind that are typical of all arthropods, from beetles to tarantulas, scorpions to centipedes. The biological affinities of the trilobite were placed beyond dispute at a stroke. These were the "natatory appendages"—Walcott evidently favoured their function as swimming limbs. Outside these legs were some other branches, three in Walcott's original reconstruction, one of which branched off the inside of the leg. The outer ones arose from a common base and were very slender, with an odd, corkscrew structure. These were the "branchial" appendages—gills for breathing, absorbing oxygen from sea water, as all arthropods must do. All in all, it was a most plausible arrangement.

Walcott had been influenced in his reconstruction by using the limbs of living crustaceans as a model. He admitted as much to his diary. "The more I study & compare with the recent crustacea, the more clearly can I see the true relations of the more fragmentary parts" (12 July 1877). Though scientists offer a convincing pretence of independent observation, even the best of them are inevitably carried forward by preconceptions—they have plausible scenarios lodged in the back of their minds, and such provisional versions of the truth are based upon previous reading and experience. Walcott carried with him the notion of trilobites as related to crustaceans, an idea that was "in the air"—recall Thomas Hardy's description, "one of the primitive crustacea called trilobites": that was published in 1873. Walcott's provisional facts* and Hardy's fictional use of the same presumptions converged in the same decade on opposite sides of the Atlantic. Who knows if these utterly different intelligences had once fingered the same textbooks?

---

*Before ten years had passed Walcott had favoured an alternative view that trilobites were related to the horseshoe crab, *Limulus*.

Trilobite limbs were to leap clearly into three dimensions thanks to another discovery in New York State. The relevant specimens were from rocks which looked very different from the limestones that Walcott studied. They were dark shales, as black as the formal hats that gentlemen wore for their visits in polite society. Cleaved with a hammer they could be split into thin sheets. The Utica Shale was exposed in a quarry in the vicinity of the town of Rome (this part of New York State was, and still is, a veritable classical gazetteer). Specimens of a trilobite called *Triarthrus* a centimetre or so long abounded on certain levels within the Utica Shale. A fossiliferous slab could look as if so many large woodlice had crawled into the rock and died there. On one such piece a graduate student, W. D. Matthew, noticed something projecting from the front of the cephalon of the trilobite; it was like two threads of gold, gently curved. Under a hand lens the delicate threads tapered, and were divided into segments. They were antennae! These were appendages that Walcott had not recognized in his sections. Antennae are the advance guard of the arthropod body: nose and fingers combined, they apprehend their environment with exquisite sensitivity. The description "feelers" that we used as children hardly does them justice. Trilobites had eyes to see, and antennae to smell and touch; already, they begin to seem less "primitive."

Professor Beecher of Columbia College was quick to appreciate the importance of the find. The source of the magical trilobites became known as "Beecher's trilobite bed," and by the end of the same year, 1893, he had recorded the new information in print. He then found the rest of the limbs on the underside of *Triarthrus,* but rather than obscure sections, here were whole legs laid out, outlined in gold. What had originally attracted attention to the antennae was a gilded film that had replaced the original delicate cuticle of the trilobite: not gold itself, but fool's gold, iron pyrites. What a miracle of preservation—as if a preternaturally delicate hand had sprayed a shiny preservative over the most evanescent portions of the anatomy, down to the last fine spine. Some specimens preserved on their backs look pinned down almost as if for a biologist's inspection. These were the specimens that at last resolved the ambiguities of trilobite anatomy: the end of a journey towards comprehension that began with Edward Lhwyd's "flatfish." Now it could be seen that slender legs, composed of many joints, lay underneath another appendage. There were not two entirely separate appendages, as in Walcott's original sketch, but a far simpler arrangement. An upper appendage was attached to the jointed leg at or close to its base (the limb was therefore described as *biramous:* with two rami, or branches), and comprised a fine, feather-like brush of filaments. Every segment along the length of the animal carried a similar pair of branched appendages. The trilobite underside was thus composed of a series of repeated units of rather similar design—one per segment, a dozen or more under the thorax. Even the segments on the pygidium seemed to have a pair of appendages of like form, getting smaller and smaller towards the rear of the animal. Under the headshield there were three pairs of similar paired limbs, in front of which were a pair of antennae—unbranched and curved slightly outwards. Professor Beecher made a series of models of *Triarthrus* from the underside which set his vision of the anatomical truth in sculptural form. I have one of these models in front of me: it is a plaster-of-Paris creation about twice life-size. Such is its verisimilitude that I have more than once been asked to display it as "the real thing."

It is *not* the real thing, of course. About every thirty years since Beecher's original report, other eyes have looked at the Utica *Triarthrus* and noticed different things. The latest of these observers was the meticulous Harry Whittington, who in the early 80s employed a young researcher called John Almond delicately to excavate the pyritized limbs with an air abrasive—a tool that blows fine powder on to the surface of the shale. The powder is harder than the shale but softer than the limb material, which is therefore revealed without damage (or so the theory goes). Whittington and Almond were able to see clearly the structure of the jointed walking limb, as explicit as a lobster limb laid out on a plate—they even found little bristles at the tip of the "feet." The upper limb, the delicately combed one, was thought to be a gill, a respiratory organ. It extended almost horizontally under the pleural lobe; indeed, successive combs may have overlapped one another. Many arthropods have a pleated "lung" through which oxygen dissolved in water can be absorbed over the folded surface. In their modern fashion, Whittington and Almond confirmed Walcott's original interpretation of "branchial appendages" more than a hundred years after his first tentative sketches. But they noticed some new things, too: for example, that the base of the walking legs—and some of the other joints—were equipped with surprisingly stout spines. *Triarthrus* was an altogether pricklier proposition than Beecher would have had us believe.

There is no final truth in palaeontology. Every new observer brings something of his or her own: a new technique, a new intelligence, even new mistakes. The past mutates. The scientist is on a perpetual journey into a past that can never be fully known, and there is no end to the quest for knowledge. John Dryden put it well:

> *In this wilde Maze their vain Endeavours end:*
> *How can the less the Greater comprehend?*
> *Or finite reason reach Infinity?*

There is always a new thought, or a new observation. Those who crave the certainty of absolute knowledge had better not embark on the endeavour, for they will be frustrated. Every cherished truth will be revised by those that follow. Real advances are made, of course, but how do we know that the end of the path has been reached? This applies to trilobites as much as to the fundamental particles of matter. Professor Beecher thought he had seen the truth of trilobite legs,

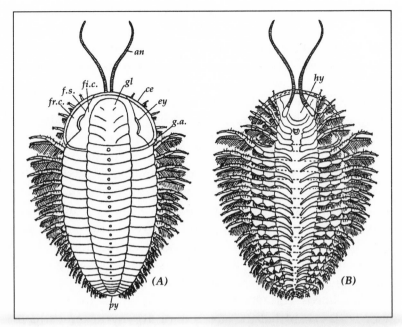

Professor Beecher's reconstruction of the Ordovician trilobite *Triarthrus* from the underside, showing the branched limbs and the antennae. There have been many changes since, but the basic facts are here. The head is now known to include three pairs of appendages (plus antennae). Right, the ventral surface of *Triarthrus*, showing the pyritized limbs prepared by John Almond from "Beecher's Trilobite Bed."

an–antenna
ey–eye
gl–glabella
g.a.–genal angle
hy–hypostome
f.s.–facial suture
fr.c.–free cheek
fi.c.–fixed cheek
py–pygidium

and sought to preserve that truth in his models—to fix his authority—but work that followed was to produce a new truth.

The latest visitor to Beecher's trilobite bed was Derek Briggs. A hundred years after Beecher there were still new steps to take along the path of discovery in the same quarry—although it had to be re-excavated. Derek, like myself a student of Harry Whittington, had become fascinated by the most obvious but mysterious aspect of the fossils, the preservation of their limbs by iron pyrites. He wished to understand why this one bed of rock produced wonderful specimens, with their golden limbs, where most rock formations yielded only empty shells. After all, the first slab I ever cracked open in South Wales was a dark mud-rock superficially like Beecher's bed—but where were the legs, and why no antennae? It was clear that the covering of iron pyrites must have happened very quickly, otherwise the legs would have decayed away. The animals themselves probably died suddenly, and were then protected from the activities of scavengers, which normally pick carcases apart even before the decay bacteria can get to work. There were clearly special conditions on the Ordovician sea-floor in New York State at this one moment in geological time.

Close study of the shales showed that the sea-floor of the time was very low in oxygen, and beneath the surface of the soft mud lacked it completely. Such inhospitable environments, termed anaerobic, are well-known today. They are capable of supporting very little in the way of animal life, but there are special kinds of bacteria which revel in them. In the almost complete absence of oxygen, they have developed special methods to extract energy from their own biochemical reactions. High concentrations of iron and sulphur are typical; the bacteria use the sulphur in their metabolism. It is very likely that it was the activities of millions of these tiny bacteria that promoted the deposition of iron on the limbs. Derek is currently trying to understand the process by devising experiments to reproduce artificially what nature did millions of years ago. It is proving complicated, but enough is known to visualize the scene. Poor *Triarthrus* was overwhelmed and died, possibly poisoned by a sudden drop in oxygen—nothing can breathe in its complete absence. No scavengers dared struggle through the lifeless water. The limp limbs were enfolded in the soft embrace of the mud, where only bacteria thrived in a soup rich in iron and sulphur. They painted the surface of the limbs before decay could obliterate them. So preserved in iron pyrites, casts of the limbs defied time.

Now it is my turn briefly to enter the story. One mystery about *Triarthrus* remained unsolved. If this Ordovician seafloor was so inimical to life how could it be that so many of these little trilobites seemed so happy to live there together, at least until they were killed? Very often they are found almost alone, with no other trilobite species (or indeed other fossils) alongside them. Nor was this unique. *Triarthrus* is only the youngest of a whole trilobite family, known as Olenidae, with a history going back more than 50 million years into the Cambrian strata. I had come across Olenidae first on the bleak shores of Spitsbergen, where they abounded in similar, but older, Ordovician black rocks, which were deposited on a seafloor where nothing else could thrive. These rocks were so sulphurous that they stank of rotten eggs when you broke them under your geological hammer. It was obvious that these trilobites had some special secret that enabled them to prosper in a home redolent of brimstone. *Triarthrus* was there, but so also were some related—and hitherto undiscovered—larger trilobites: one of these I christened *Cloacaspis*, after the Roman sewer *cloaca*, for reasons which will by now be obvious. (The original, constructed by Tarquinius Priscus, conducted the filth from the streets of Rome into the Tiber.) I was to see similar, still older rocks around the Norwegian capital, Oslo, where the Cambrian ancestor of all these trilobites, *Olenus* itself, abounded. The smelly nodules which yielded their remains are probably the only noisome thing in this most orderly and spotless city. *Olenus* was one of the earliest known trilobites, named by the pioneer Scandinavian geologist J. W. Dalman in 1827 after the husband of Lethaea. Together husband and wife were turned to stone by the gods—the most appropriate classical names were grabbed by early palaeontologists! All these olenid trilobites flourished in sulphurous muds rich in iron, in a habitat in which there was very little oxygen—conditions which effectively banned all the competition.

It is only in the last few years that animals adapted to a similar habitat have been investigated in detail. Nature is adept at making a virtue of necessity, by turning hardship into opportunity. In stinking mud today there are types of clams that grow special bacteria in their gills. The bacteria process sulphides, and the clams can maintain just enough oxygen to allow the bacteria to do their work—too much oxygen would actually oxidize the sulphur compounds that provide the bacteria with their food. They are thus animals on the edge. They live only in low-oxygen habitats, where below the soft sediment surface there is no oxygen at all and sulphur compounds abound. The special bacteria are known as colourless sulphur bacteria, and they require modern microscopical and molecular techniques for their study. Neither Beecher nor Walcott could have had a clue as to their existence. Clams can absorb nutrients from the bacteria directly, but other creatures that "cultivate" them use them as food.* When I studied olenids I, too, had concluded that they lived in a low-oxygen, marine environment stinking of sulphides. Then I came across records of the living symbionts that grew their own sulphur bacteria, and discovered under what conditions they thrived. It was a grand moment. Suddenly it was possible to make sense of so many features of Olenidae. It was now obvious

---

* For those who like polysyllabic descriptions to impress people I should say that these animals are correctly referred to as chemoautotrophic symbionts.

why they lived in numbers together to the exclusion of other species, and in company with such unpleasant rocks: this was their speciality. They had long bodies with large numbers of segments in the thorax—all the more space to grow bacteria. They may even have grown them along the long "filaments" of the gill branch, like the living clams. Safe from predators they had very thin shells. There was iron available to replace the limbs. All the facts hung together: these particular trilobites were the first known animals to live symbiotically with sulphur bacteria.

The Utica Shale was not unique in its exquisite preservation of trilobite limbs. In Germany, to either side of the River Mosel and in the adjacent parts of the Rhineland, another dark slate comes to the surface. The Hunsrück Slate has been used for roofing since medieval times, and by the middle years of the nineteenth century there were many productive quarries. Even today, an open cast working at Bundenbach still employs thirty slate splitters. In geological age this slate lies between the Utica Shale and the Carboniferous with which this book began; it is early Devonian (about 390 million years). The slates were squeezed as part of the complex Hercynian earth movements, the same convulsion that affected the slates that made wild Pentargon Cliff on the Cornish coast. At certain levels in the Hunsrück Slate fossils are replaced by iron pyrites, just as they were at Rome, New York. But the circumstances differ. In Hunsrück an entire, rich marine fauna is pyritized: starfish, sea lilies, worms, fish. Creatures with soft bodies were caught by surprise, as in a snapshot. If this book had been about starfish I could have filled pages with a celebration of the Hunsrück wonders. There was nothing restricted or peculiar about this Devonian sea-floor: it was bursting with life, and there was presumably plenty of oxygen. Trilobites were just one of many kinds of arthropods that crawled around on the soft Devonian muds, although they were the most abundant. It is now thought that the whole sea-floor was overwhelmed by occasional muddy slurries that

The legs of the Devonian trilobite *Phacops* preserved in iron pyrites from the Hunsrück Slate, Germany. (Photograph courtesy Prof. W. Haas.)

were sufficiently charged with iron to effect the pyritization of the fossils. Professor Wilhelm Stürmer perfected a technique of taking x-ray photographs of these buried animals. Although it is possible carefully to dig out the fossils, as John Almond did from the Utica Shale, how much better to peep inside the solid rock to photograph the buried animal, which William Conrad Röntgen made visible by his discovery of x-rays in 1895. The iron pyrites is more opaque to x-rays than the surrounding slate, and the fossils are outlined on their radiographs as if drawn by a deft artist in a soft, dark pencil.

They have a ghostly quality; one might imagine they had been called up from the past by incantation rather than by science.

The commonest of the trilobites in the Hunsrück Slate is an animal some centimetres long called *Phacops*. I have in front of me a wonderful x-ray of this animal (fig. 6) in which the appendages are shown clearly, but superimposed by the x-rays, as if they were in a state of fluttering movement, like the walking figures in the Futurist paintings of Umberto Boccione. They are the nearest thing we will ever see to a living trilobite, but we see it through a photographic plate, darkly. *Phacops* is not closely related to *Triarthrus*, and it is surprising to find that their appendages were in many ways alike: similar antennae, similar paired limbs on every segment. The x-rays reveal the fine tips of the gill branch better than a pin ever could. Here is proof that fossils could record something more delicate than lace, and as fugitive as a cobweb.

As more and more trilobites with preserved appendages were discovered, most of them were found to have limbs of similar kinds along the length of the animal—and each limb comprised a paired walking leg and gill branch. Trilobites did not develop the specialization of individual limbs which happened in many other kinds of arthropods. Think of the nutcrackers of the lobster, or the extensible sucking pad of the fly. Instead, most trilobites retained a comparatively unspecialized locomotion; it was the shell that evolved into an array of fantastical shapes. Carnival floats flaunt colourful and extravagant paraphernalia, and it comes as a surprise when the dressing is removed—beneath is a humdrum Ford. We have laid bare the workings of the trilobite: what lies under the chassis is no longer mysterious. We are ready to envisage a parade of trilobites walking past on their paired limbs: and it will be as odd a parade as any carnival could offer. Some smooth as eggs, others spiky as mines; giants and dwarfs; goggling popeyed popinjays; blind grovellers; many flat as pancakes, yet others puffy as profiteroles. There are thousands of species. They are so prolific that they have been dubbed "the beetles of the Palaeozoic," and beetles are the most bewilderingly diverse of living organisms; biologists are left breathless trying to calculate just how many species there may be. Nor are we anywhere near knowing how many kinds of trilobites still lurk undiscovered in the rocks. Our trilobite march-past can only be a selection of a selection. It will cover 300 million years of history in a page or so. A glance at the illustrations will give a better idea of the extraordinary variety of form that the trilobites achieved. The parade will be in approximately geological order, with the oldest first. How they evolved into such a variety will be the subject of a subsequent chapter, and the trilobites described here nearly all appear as characters elsewhere in this book. To me, all the creatures named are as familiar as kin.

First comes *Olenellus* (fig. 10), commonest of the earliest Cambrian trilobites (535 Ma). It was discovered in the mid-nineteenth century by the pioneer New York State palaeontologist James Hall, and has subsequently been found very widely, including as far away as Scotland. Although it is so ancient, its large and long cephalon already carries a pair of extended eyes. The widest part of the animal is at the head end where there are prominent spines at either corner, behind which the body tapers gradually backwards along a thorax comprising many, rather flat segments with prominently spiny tips. One of the thoracic segments, near the front, is much more strongly developed than the rest, so that the pleural spines project beyond the rest of the body. Then there is a long spine rising from the middle of the axis of the thorax, near the back end of the animal, behind which the segments are very tiny, and the pygidium is really minute. Somehow this *looks* like a primitive trilobite. It has not yet developed the sutures crossing the headshield that helped its relatives during moulting. The tapering glabella is very clearly divided by furrows into segments, and its front is almost round, like a boss.

Olenellus is followed by a giant, the size of a large lobster. It is a rapid mover, striding purposefully in pursuit of smaller fry, which it can spot with its glistening eyes. This is *Paradoxides* (see p. 224), which has already been introduced as having a name to match its oddities. It was originally discovered in Sweden in the early nineteenth century; now it is known very widely. It too has many thoracic segments, but lacks one considerably larger than the rest. The genal spines are fearsome, extending backwards like a pair of swords. Pleural spines at the back end of the animal extend backwards beyond the tail like the kind of drooping moustaches sported by the bad guys in westerns. The tail, although a little larger than that of *Olenellus*, still does not amount to much. But the furrowed glabella is swollen—the whole thing expands forwards, and beneath lies a stomach which must also have been enlarged, probably to engorge prey. *Paradoxides* is mid-Cambrian in age, fifteen million years younger than *Olenellus;* still early days, one might say, but *Paradoxides* clearly meant business.

Now there is a flickering swarm—are these trilobites, too? They look like tiny animated beans, just a few millimetres long. This is a fly-past (or swim-past) rather than a parade, for these little animals are sculling through the water like so many water fleas. They are so small that you to have to squint hard to see how different they are from their other Cambrian relatives. Some of them seem to be rolled up tightly. They are as different from *Paradoxides* as can be imagined, and not only with regard to size, for these animals have very few, perfectly hinged thoracic segments—in fact, only two, which are bluntly tipped as if chopped off with a tiny scalpel. And it is hard to tell head from tail: both are equally large, and there is no sign of eyes. So these are little blind trilobites which have turned into something totally different from the crawling spectator of Mr. Knight's clifftop predicament. They are strange miniatures, specialized, sophisticated, and so successful that when there was plenty of plankton on which to feed they must have turned the late Cambrian (505 Ma) seas dark with their numbers. They are found in every continent in rocks of the right age. *Agnostus* is the name for these little enigmas—and how appropriate!—and the species that is so common in our parade is called *Agnostus pisiformis* (fig. 11)—literally, the "pea-like unknown one." The genus name was given to *Agnostus* in 1822 by Brongniart, who will be recalled as having named the "Flatfish" from Llandeilo as a good trilobite. I have handled limestones from Sweden which are made of almost nothing *but* this tiny agnostid trilobite—rocks that can look like petrified pea-soup, a cobble of knobbles. Curiouser and curiouser.

Here comes another stately animal. It is the size and shape of a small silver salver, rather convex and smooth, and, like *Agnostus*, it has a pygidium as large as the headshield. But there the resemblance ends, for this trilobite has eight thoracic segments, each perfectly faceted so that rolling-up can be effortlessly achieved. Its eyes are prominent, shaped like crescent moons and elevated on stalks so that they resemble a pair of up-periscopes perched atop the head. The glabella is not so clearly marked as in *Paradoxides*, neither are its furrows so deep. Nor are there are any genal spines, so that this trilobite looks polished and rounded off at the corners, built for smooth ambulation. One of these animals has partially buried itself in the soft mud, where it can be recognized only by a vague disturbance in the sediment surface, and by its eyes projecting above it, unblinking in their vigilance. *Isotelus* (fig. 12) is a more convex version of Lhwyd's *Ogygiocarella* from Llandeilo in South Wales—and indeed it is one of its close relatives, and of similar Ordovician age (470 Ma).

Isotelus dwarfs some of its contemporaries, one of which is a little medallion-like creature with a puffy head, and almost flat thorax with six segments and a perfectly triangular tail. These trilobites have enormously long genal spines—far longer than the rest of the body, so that the animal is supported on them like a sled upon its runners. The glabella is higher than the rest of the animal, and inflated like a pear; it is hard to see any evidence of eyes, so this was probably another

blind trilobite. Most extraordinary of all, the head was surrounded by a border full of perforations, like a colander. The
pitted fringe lay at the front of the head in the manner of a
halo—and the pits were not just scattered willy-nilly. Instead,
they were organized clearly into rows and lines, each species
having its own particular arrangement. Each has the size and
perfection of a coin, minted to a design as precise as a dynastic
intaglio. On the underside, rather feeble little limbs propelled the little medallion from spot to spot: these species
did not leave the safety of the sea-floor for long. The name *Trinucleus* (fig. 13) is too obvious to require any explanation
from a Latin thesaurus; it was donated to these peculiar animals by Sir Roderick Murchison in 1839 after he had made his
seminal Welsh traverses (1833–7). Modern study has shown
that the fringe around the head is matched closely by a modified lower layer of shell, a "lower lamella" on which every
pit is opposed by a corresponding tubercle. The fringe is a
complicated piece of natural engineering, a double-up plate
punctured by little tubes. This animal must have been as specialized in its Ordovician fashion as anything in the modern
ocean. But how it lived remains an enigma: five generations of
palaeontologists have studied these perfect little animals and
wondered. *Trinucleus* itself is confined to Wales, but its close
relatives are found worldwide.

Now some swimmers scull into view. Here are trilobites
that are extraordinarily gifted with eyes. They bulge prominently, as if they suffered some thyroidal condition. Almost
the whole of the side part of the headshield has become a
great, inflated visual surface: the free cheeks have become
converted into nothing but eyes. The honeycomb pattern of
the lenses is as clear as in any dragonfly. Even more bizarrely,
the eyes have actually fused together at the front of the animal, so that effectively there is just one huge visual organ, or
headlamp. This is *Cyclopyge*, discovered originally by the
Bohemian palaeontologist Joachim Barrande in 1845. (A very
close relative is shown at fig. 15.) The name is derived from

the Cyclops, mythical one-eyed giants of classical Thrace. Our trilobites are not giants except in the optical department; overall, they are the size of large bees. But in the matter of eyes, what a wonder! The rest of the creature is smoothed out, and compact. It is hard to see exactly where the glabella is on this trilobite: it remains as a flat area between the eyes; there are six strong thoracic segments, each one well-articulated to its neighbours. This animal was built for swimming. The pygidium was nearly semicircular, and had a short axis. While *Trinucleus* grovelled on the sea-floor, *Cyclopyge* cruised above it.

Illaenus is about as convex as trilobites get: the whole animal resembles an armoured carrier, with all its edges rounded off. Its eyes were rather small and set high up on a head that sloped steeply down to its margin. Glabella and thorax merged smoothly into one another, and the thoracic pleurae hung down steeply. The tank-like profile continued into a large semicircular tailshield, which was also nearly smoothed out, so that you have difficulty guessing how many segments went into it. When *Illaenus* rolled up it attained a nearly perfect sphere, unbreachable by any predator. It was an armadillo among trilobites. (*Bumastus,* a close relative, is shown on fig. 3.) It is not hard to imagine a predatory Ordovician or Silurian contemporary of *Illaenus* trying in vain to prise open one of these tight capsules; and all the while the crystal eyes gazed on to register the frustration of its foe, as the animal bided its time until it could unroll and scurry off to safety. As with so many trilobites, *Illaenus* was first discovered in Sweden in the early nineteenth century, and has since been recognized on nearly every continent.

Calymene is regarded by many as the typical trilobite. This is probably for no better reason than that it has appeared in so many textbooks as the first example for students to learn. It is one of the commonest trilobites in Silurian strata (about 425 Ma) and was found in the classical Wenlock district, where are exposed some of the first British Lower Palaeozoic rocks to be fully investigated. A. E. Housman's Wenlock Edge is a bluff of

Wenlock Limestone, commanding a view westwards to Wales, where heads of fossil corals weather out slowly under the gentle wash of the Shropshire rain. In the town of Dudley, Worcestershire, quarries which were active in the eighteenth and nineteenth centuries yielded hundreds and hundreds of beautifully preserved examples of Blumenbach's Calymene (*Calymene blumenbachii*). Any collection worth its salt has one or two of these specimens. They are singularly satisfying things, palm-sized, plump, emanating an undeniable, primal charm. One of the objects I have to lock away in the collections I look after is a marvellous, gold-mounted *Calymene* brooch (fig. 17) which once must have made a most arresting conversation piece when pinned on the bosom of its former owner. The "Dudley locust" appears on the coat-of-arms of that town (and surely those who named it knew an arthropod when they saw one, even though "locust" is a little approximate). The enthusiastic curators of the local museum want to build a great education centre in the old quarries in the form of a huge *Calymene* where visitors can dine under the glabella and learn history under the thorax. I am all for it. The convex trilobite had a tapering glabella carrying very deep furrows; free cheeks like two-thirds of a circle; twelve thoracic segments with a very prominent axis; and a neat downsloping tail smaller than the head, which tucked underneath the cephalic rim when the animals enroled. I like to hand round rolled-up *Calymenes* to schoolchildren so that they can weigh more than 400 million years of history in their hands. Such an engagement with the real object is worth a dozen videos. A sense of wonder is not to be bought over the counter at the superstore. Nor is it something which can be wheeled out of a corner cupboard at the behest of some curriculum or other; instead, it steals up on the child unexpectedly.

Radiaspis is a hymn to prickles. It is smaller than *Calymene*, but makes up for it by sprouting spines everywhere. All around the front of the cephalon there is a comb of spines; there are genal spines; there is not one, but two spines arising from the tips of each of the flattish thoracic segments; and there are long, elegant spines arrayed around the pygidium. You have to look hard to recognize that the stalked eyes are not yet another pair of spines. *Radiaspis* is an odontopleurid trilobite, which means "toothy pleura" and you can see why. The glabella is divided into curious round lobes; there are further spines on the axis of the thorax. You instinctively know that this was another specialized animal, because it evokes that same sense of oddity you experience when you see a seahorse for the first time, or a long-eared bat. You feel suddenly in awe of the richness of the natural world. Odd it may be, but the odontopleurid design (see fig. 32) was singularly successful, lasting from the Ordovician until the Devonian (500–370 Ma) and generating hundreds of different species, each bedecked with a different arrangement of the spines. The related genus *Dicranurus* is illustrated on the cover of this book carrying its ramshorn-like spines at the back of the head.

And still they come. We have already met *Phacops* (fig. 18), a large Devonian trilobite having special eyes with enormous lenses, through which we shall see the ancient world with special clarity in the next chapter. The first *Phacops* specimens were discovered in Germany in the 1820s; then they were found in Britain, France and North America. The carapace is covered with coarse, warty tubercles. As I write I am running my hand lightly over the surface of a large *Phacops* from Morocco. These particular trilobites have been commonly available on the market since about 1985, and most are rather roughly dug out from their limestone bed, so that they almost have the look of a sculpture. My own specimen feels rough, like one of those old-fashioned dill cucumbers; my fingers feel out its eleven thoracic segments. This trilobite is so well-defined by ridge and furrow it can be read like Braille. The glabella expands forwards into a triangle; the pygidium is deeply segmented. One of the common species is *Phacops rana*—the describer must have had the warty "skin" in mind, *rana* being the Latin for a frog. Some specimens from the

1. Probably the first trilobite to be recorded in a scientific journal, Dr. Lhwyd's "flatfish," now known as *Ogygiocarella debuchii*, from the Ordovician rocks of South Wales, near the town of Llandeilo. Eight thoracic segments and a large pygidium, and big crescentic eyes. Natural size.

2. Silicified trilobite headshields, dissolved out of limestone using acid. Harry Whittington's preparation of the Ordovician trilobite *Ceraurus* from the Ordovician of Virginia. This headshield from the underside (b) shows the hypostome in position, and viewed from the front (c) you can see how it bulges downwards beneath the glabella to house the vital head organs of the trilobite.

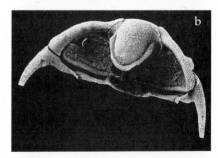

3. *Bumastus,* a 10-cm-long trilobite in which many of the features have become smoothed out, or effaced. You can hardly discern the glabella, though the eyes are quite prominent. The whole animal is very armadillo-like. Silurian, Shropshire, UK.

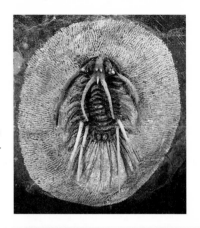

4. *Radiaspis,* a fantastically spiny trilobite from the Devonian rocks of Morocco. Delicate spines originating from the thorax curve back over the spiny pygidium; another pair of spines come backwards from the "neck" region.

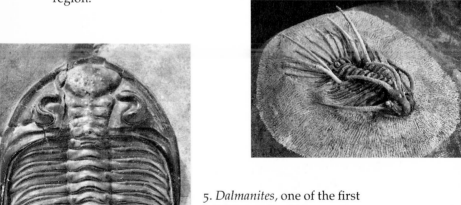

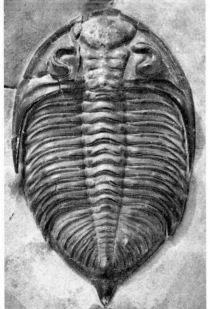

5. *Dalmanites,* one of the first trilobites to be discovered, abounds in the Silurian rocks of Europe, often about 10 cm long. A short spine at the back of the tail is characteristic. This specimen is from Shropshire, UK.

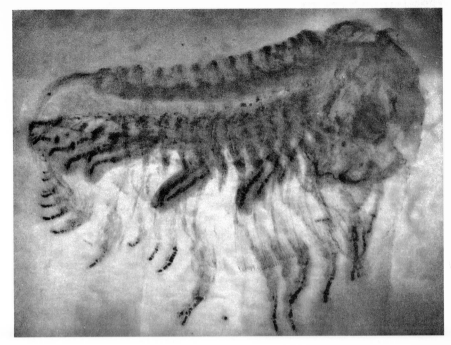

6. X-ray of the Devonian trilobite *Phacops* from the Hunsrück Shale of Germany. The legs appear ghost-like and the fringes of the gill branches show at the edge.

7. *Olenoides serratus* from the Cambrian Burgess Shale. Details of the limbs taken from Harry Whittington's incomparable photographs: the thoracic segments are on the left; the strongly spinose walking legs extend below.

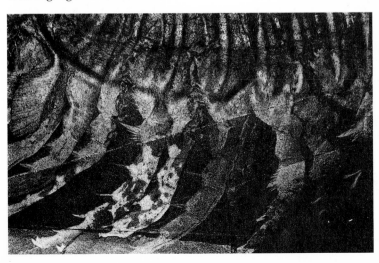

8. An olenid trilobite, *Hypermecaspis*, approximately life size, showing small headshield and long thorax with numerous segments. This specimen is preserved upside down so that it shows the hypostome (covering the stomach) more or less in life position. Ordovician, Bolivia.

9. A "graveyard" of the olenid trilobite (see p. 69) *Leptoplastides* from the early Ordovician shales of Shropshire, UK. You can see individuals of more than one size on this single piece of rock, and about equal numbers are right way up or upside down. Most of these individuals are a centimetre or two long.

10. *Olenellus,* one of the oldest trilobites from the Lower Cambrian rocks; the example illustrated is from Pennsylvania. In spite of its antiquity it has clearly developed, large crescent eyes. Notice how the third thoracic segment is larger than the rest. The pygidium is minute—partly concealed beneath a spine towards the rear of the thorax. *Olenellus* is often as long as 10 cm.

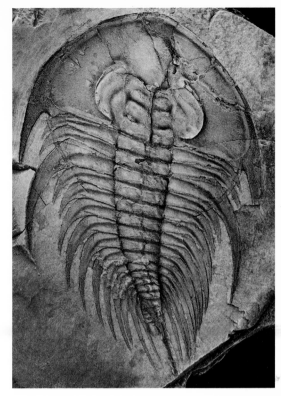

11. The tiny, blind *Agnostus pisiformis,* a few millimetres long at most, from the late Cambrian strata of England. Cephalon and pygidium are almost identical in shape, and there are only two thoracic segments.

12. The elegant *Isotelus* from the Ordovician of New York State. Eight thoracic segments show this to be a relative of *Ogygiocarella*. The outline of the head and tail match closely—a match that was used during enrolment. The curved eyes stand proud of the head. About 10 cm long.

13. The blind, medallion-like trilobite *Trinucleus fimbriatus,* with its remarkable "fringe," the function of which remains debatable. This is a moult—the genal spines are missing. Specimens are a few centimetres long. Ordovician of Wales.

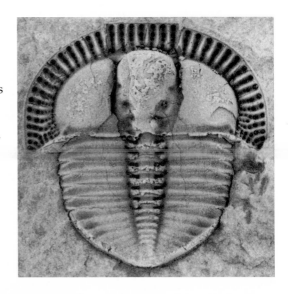

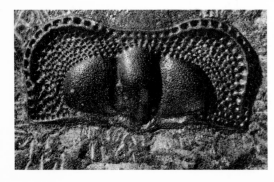

14. The head of an Ordovician trilobite related to *Trinucleus*, *Protolloydolithus*, recovered from a borehole in eastern England. This beautiful trilobite has a symmetrical "fringe" perforated by a hundred pits.

15. The head of the giant-eyed trilobite *Pricyclopyge*, which swam in the deeper waters which covered southern Wales in the Ordovician period; *Pricyclopyge* is usually 3–5 cm long.

16. Derek Siveter's photographs of the Silurian trilobite *Calymene* from Gotland, Sweden, in four different views. The forward view shows how the tail is perfectly shaped to tuck up inside the head-shield.

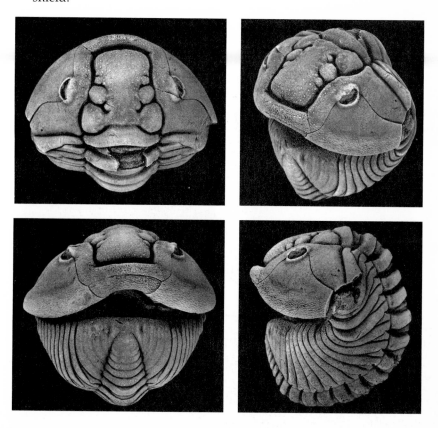

17. A unique antique gold brooch with the Silurian trilobite *Calymene* as the centrepiece.

18. A lateral view of a perfectly preserved enroled *Phacops* from the Devonian limestones of Morocco. Very similar species occur in North America and Europe and the Far East. The glabella carries great tubercles. The eye with its spherical and slightly sunken lenses is shown well in this example.

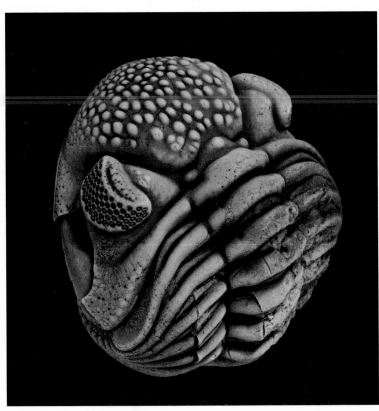

Silica Shale in Ohio weather out from the white shale as if moulded in burnished pewter. They are found in clusters, and at least one observer thought that they gathered to mate before being overwhelmed by disaster. If so, a moment of pro-creation would have become the moment of preservation.

There are many more trilobites which scuttle past so swiftly that we can see only one or two salient details before they pass from view. Here is *Crotalocephalus* (fig. 19) with a tail like a cat's claw carrying a few large, curved spines; then *Dal-manites* with its single spine extending from the back of the pygidium like a marlin spike. Now comes *Scutellum* (fig. 21), flat as a flounder, but with a huge pygidium shaped like a ribbed fan—one flip and it is past us. A gigantic *Lichas* is nearly as flat, but with the glabella blown up, like so many bubbles, and with a fluted pygidium with a coarse saw-edge. Nearby, there are some tiny trilobites buried in the soft mud, probably diminutive blind grubbers called *Shumardia* (p. 231). There follow more spiny monstrosities—*Comura* (fig. 34), with so many vertical spines it looks like a fakir's nightmare. And so the parade passes on and on.

The very last trilobite is a smallish one, *Phillipsia* (close rel-ative *Griffithides*, fig. 23). This animal is named after John Phillips, whose *Illustrations of the Geology of Yorkshire* (1836) earned him the right to be so immortalized in stone. At first glance this new trilobite does not seem such a remarkable ani-mal compared with some of the phantasmagoria which have flickered before us. But it would have been a relative of this species which Thomas Hardy placed into Beeny Cliff in North Cornwall; it was a Carboniferous (330 Ma) crawler with large crescentic eyes, and the whole shell dotted with prominent tubercles as if it were suffering from an attack of Palaeozoic measles. The tapering glabella seems as badly infected as all the other body parts. The tailshield is very large, and deeply furrowed. Maybe Hardy had seen illustrations of this very animal drawn by Phillips from Yorkshire, and saved the image in his eclectic mind. There is no particular feature that marks it out as a survivor, but that is what it was, for several relatives of *Phillipsia* were the last trilobites of all. They lasted into the Permian period (260 Ma) before they and their kind became extinct. At last, the 300–million-year-long parade has come to an end.

It should be clear by now how a lifetime can be spent trying to catch a few moments of this parade. There is so much history to learn, and only a handful of shells to reveal it to us. For every animal that clearly comes into sight another dozen escape, or leave only their footprints in the soft mud. A puzzled fellow commuter on the train once asked me how I could go to the office day after day to study a trilobite. I think he believed that there was only *one* trilobite, rather like the Mona Lisa, and that my day was spent contemplating it and making up new theories about its enigmatic smile. I had to explain that my work was more like attending to an almost infinite system of galleries hung with Mona Lisas, and that often all we had was the smile. And every time the end of one gallery was reached, there was another gallery beyond still to explore, and further again another . . . and hardly ever the legs.

There are some curious advantages in belonging to a small scientific world like the circle of trilobite specialists. You know almost everybody: it is like an extended family. As in all families there are feuds and disagreements, but kin loyalty wins out in the end. As in all families there is an acute awareness of family history. Charles Doolittle Walcott is regarded as one would a famous ancestor; we suffer with those that went mad, like poor John Salter, or died under Nazi persecution, like the tragic Rudolf Kaufmann. There is a solidarity that transcends generations and national barriers. It does not matter where you go; if there is a "trilobite person" there you will find a friendly face at the airport, and before long you will be swapping the names of fossils like unofficial passports. When I arrived at Almaty airport in Kazakhstan in 1996, in the middle of winter, I stood outside the rather large shed which passed as the International Terminal. Sleek limousines pulled up—

but they were not for me, they were for the new post-Soviet generation of businessmen, intent on buying and selling oil or minerals, or, for all I know, their grandmothers. As I loitered, now on my own and feeling a little disconsolate, my breath coiling up into the frosty night air, I spotted in the distance an ancient Trabant. Stuttering along like a jalopy in a Pluto cartoon, scattering rusty flakes like dandruff, it wheezed to a halt beside me. Out of the window leaned the beaming face, gold teeth flashing, of my Kazakhstan colleague Mikhail Apollonov. "I have arrived!" he stated grandly, if unnecessarily. Within minutes we were swapping trilobite intimacies and gossip.

I V

Crystal Eyes

It might seem hardly worth questioning the idea that the world is made for seeing, or that eyes are consequent upon the undeniable fact that there is so much to be *seen*. Yet think for a moment and the inevitability of vision is much more uncertain. The world is full of other signals that may be used to describe it: there are smells, chemical signals both subtle and ubiquitous, and touch is as sensitive to shape as sight—more so, because it cannot be misled by *trompe l'oeil* or by camouflage. Imagine a world in which the eye had never developed—not the eye of insect, nor of fish, nor of mammal, nor yet Mankind. It is easy to conceive of the other senses having taken over the comprehension of their surroundings. It would be a world of palpation, of feelers, a world in which caresses would have rendered glances superfluous. The twitching and waving of antennae would accompany every action. It is not difficult to imagine that a different evolutionary course would have selected those organs most delicately attuned to the passing molecule: even now we know of moths so sensitive to the pheromones of the opposite sex that the most evanescent whiff of a mate can stimulate a love flight across kilometres. In a sightless world, sensitivity to such stimuli would be selected and refined: it would be a world of nuance so delicate that our gross maulings would be inconceivable.

In conscious animals this most sensory of environments would entail everywhere the language of touch and smell: beauty would be aural or tactile or olfactory. Poetry would not celebrate the unfathomable mysteries of eyes and their unplumbable depths, nor compare hair with flax, for visual similes would be redundant. Rather, the texture of skin might be the supreme erotic stimulus, or natural selection might have favoured an ever more elaborate array of perfumes and chemical attractants, which in turn would evolve a language of which we can only dream. There might be symphonies of perfume, Mozarts of musk. Novelists might construct nasal narratives, versifiers sonnets of scent. Sculpture would entail subtleties of shape that only fingers trained through hundreds of millions of years of tactile evolution could discriminate. There would be no word for "blindness."

So I don't believe that light inevitably engendered sophisticated sight, simply that that particular path was taken by life on this planet, elaborating and improving upon the simple photosensitivity of single-celled organisms. The eyes of the trilobite are tangible proof of a selection of one special branch of evolution from an array of possible alternatives—an innovation that made the world visible. Once passed, this threshold could not be forgotten, even if some animals—trilobites included—once more lost sight of the world in favour of fumblings in the dark.

Recent laboratory work has revealed the pervasive influence of the genes that control the sequence of development of the various organs as animals grow from embryo to adult. The master control is the family of HOX genes. It is astonishing to find that similar genes control the placing of the head in the locust as in the fish (or kangaroo, or human). These are genes so deeply embedded in the unconscious mind of our bodies that the memory of their origin is lost far in the Precambrian history of development of the earliest animals. We can never, ever, directly sample the genetic code of the trilobite—but we can be sure that its development was under the control of HOX genes of kinds that we could still recognize in living animals.

The argument that allows this assertion is entirely a matter of logic. Embryologists who study the cell-by-cell growth from fertilized egg to recognizable organism have been able to develop techniques of staining tissues which are responding to the bidding of particular genes. This is how they know that the development of the standard experimental insect *Drosophila* proceeds in a similar fashion to the developing embryos of vertebrates. What they have discovered is a list of instructions about how to order a body that operates whatever kind of body is being constructed. The conclusion follows that the genes that control this sequence of development must be so ancient that they go back before the last common ancestor of insect and vertebrate. This is an evolutionary "split" of such antiquity that it must precede even the oldest trilobite, because we know for sure that trilobites were already typical arthropods (like their distant cousins—the fruit flies, *Drosophila*). So the common ancestor connecting backboned animals and joint-legged animals—an animal that must have owned the right genes in common—must lie further back in time from the first trilobite fossil. In fact, the split between the arthropod and the vertebrates is one of the deepest branches on the tree of life—it is even likely that arthropods are more closely related to snails than they are to vertebrates. We cannot readily conceive of this remote ancestry; probably we will never know what that common ancestor looked like. We can infer that it was small and soft-bodied, and it will have left no fossils. Nonetheless, its legacy still tells the developing embryo which cells shall form a head, and how the body shall be arranged from front to back. It inexorably follows a blueprint originally drawn up in the most ancient times. It is rather wonderful to imagine this distant manifesto at work on the growing trilobite, directing the brain to be enclosed within the head—and, of course, issuing instructions for the growth and development of eyes.*

For eyes are part of this ancient list of instructions. It seems that the making of an eye is the same impulse in fish or fly or man. As the cells develop in the embryo there is a point at which eyes begin to differentiate. They start as a bundle of cells, which divide, and divide again. The end product may be very different—after all, insects have compound eyes and vertebrates sophisticated, lensar eyes—but the impulse, the instruction "make eyes!," may be common to all animals. The deep language of the genes is an Esperanto of biological design which can be understood by a Babel of organisms. The deep-seated genes represent an organizing principle that predates the extraordinary proliferation of life that made the living world as rich as it is today: to understand this deep structure we have to strip away differences to the commonality of ancestry. The eyes have it.

Maybe this democracy of eyes stretches as far back as the flatworm, a little, wedge-headed organism that still abounds in moist places in soils and under stones. Many readers will know the flatworm only as a clever cipher in one of M. C. Escher's ever-retreating symmetrical designs. In his drawing, flatworm interlocks with flatworm into an infinite regress that flaunts geometrical meshing finer and finer until it reaches a kind of *reductio ad absurdum*. It is a favourite subject for posters on the walls of advanced biology classes. The flatworm has a kind of surprised look—expressed with its eyes—as well it might, being the victim of such an exercise in facile geometry. Many biologists place the flatworm (or, to be accurate, several different flatworms) close to the common ancestry of most higher animals. Hence, the ancestor of both trilobite and train driver may be a tiny, flattened invertebrate

---

*The gene controlling eye development is not a HOX gene, but a homeodomain gene called PAX6.

with minute eye spots. And the instruction that bids the eye grow in a flatworm may be the same instruction that instructs it to form in ourselves.

So when you see—with your own eyes—the eyes of the trilobite, you recognize a kinship of vision stretching across hundreds of millions of years. It's too bad that the trilobite cannot manage a conspiratorial wink. The trilobite reminds us of that moment—at least, a moment in geological terms— when the first organism developed a cell that was sensitive to light. Then the elaboration and proliferation of such cells that followed was sealed for ever into the blueprint of development for our own sight-dominated world. Who can doubt that as soon as vision became a realized possibility it must have conferred on its possessor an exceptional advantage? Food could be identified by shape alone, and the approach of enemies would blot out the sunlight. Surely there must have been a premium on seeing the world more clearly, recognizing subtler movements, that would encourage the evolution of more and better vision. Now there was a point in painting-up to attract a mate. Colour would have a purpose. The delicate deceptions of camouflage, the machinations of mimicry, the whole of Nature's palette, would follow as a logical consequence. Without that moment of vision, colour in the natural world would have been haphazard, a splash of red here, a flash of green or yellow there. Although colours are incidental properties of many biological molecules, it requires vision to recruit them into useful roles and to paint the Earth with a purpose.

As to when this happened, we know that the first trilobites in the early Cambrian already had a sophisticated visual system. The eyes of one of the oldest of all, *Fallotaspis* from Morocco, are quite large. This animal dates from approximately 540 million years ago, hence the origination of eyes must have been before that. Several of the soft-bodied animals from the Chengjiang fauna of the early Cambrian of China also had eyes—some of them on stalks. Arthropods like

Fuxhianshuia seem to have their eyes positioned far forwards, whereas trilobites, of course, carry them on top of the head, within the shield. So it is clear that there had already been a lot of evolution of variety in eyes among the jointed-legged animals almost at the inception of the Cambrian. Whether or not this happened in a burst of accelerated evolution—the Cambrian "explosion"—is a subject to which I will return. For the moment we can say that solid fossil evidence that cannot be misunderstood proves that eyes originated before—and probably well before—540 million years ago.

An estimate of when the *first* eyes might have originated in the Precambrian can be obtained in a different, indirect fashion. Recall that body fossils, actual organic remains, are rare in Precambrian rocks, and of animals with eyes there is no really solid evidence. Perhaps these animals were very small and soft-bodied: they certainly lacked the hard and readily preservable shells of their trilobite successors. We have to do without direct evidence, and infer distant events from effects that linger on even in living animals. Since we know that animals descended one from another, we would like to know when one, basal group of eyed animals "split off" on the tree of life from their eyeless relatives. The eye-bearing animals in question may have taken very different paths subsequently— to have become as unlike as a whale and a flea, or an octopus and an orangutan. This is of no concern in the question of origins. What we are interested in is dating the branch in the road map of evolution where the different routes were initiated—this is called the "divergence time." It will be important to learn when "higher" animals split off from the flatworms, carrying with them message to "make eyes!" that still operates its imperative. The divergence time can be guessed at by summing up the changes that have accumulated in the genetic code after the split occurred. Mutations happen, and such changes gather in the genetic code like bad memories stocking a guilty conscience; the effect is cumulative. These accreted mutations can provide a kind of clock, which can be reckoned in terms of millions of years if the right part of the genome is examined. There are "fast" clocks and "slow" clocks, and to try to look back into the Precambrian we need almost the slowest clocks of all, located in parts of the genome that are enormously conservative. We need to look for the genetic Collective Unconscious shared by all animals.

There are some parts of the genetic code which have proved particularly useful in trying to date distant evolutionary events. Inside every living cell there are large numbers of tiny particles called ribosomes; they are where life-giving proteins are synthesized. About 60 per cent of the ribosome is composed of ribonucleic acid (RNA). Parts of the ribosomal RNA molecule have been claimed to have just about the right degree of conservatism to measure and calibrate the changes in which we are interested—neither so "slow" as to have remained for ever unchanged, nor yet so "fast" as to have run around the clock more than once. This vital molecule is possessed by all the animals with which we might be concerned, and so the genetic changes that have accumulated over hundreds and thousands of millennia permit a common time calibration. However, the claim that RNA "clocks" are reliable is still controversial, and many of my colleagues wonder how much "noise" there is in the signal, that has nothing to do with the ticking of geological time.

Over the last ten years or so (I write in 1999) there has been a large number of estimates of divergence times based on a variety of genes and different "bits" of the RNA molecule. Recently, as the enormous, cryptic library of the DNA molecule itself has gradually been decoded, other evidence has been brought to bear on the question of divergence times. Some of those genes which code proteins, and the DNA found in the mitochondria of the cell, have been used as "clocks" to check on ancient genetic legacy. What convinces me that there is some truth in estimates of deep Precambrian divergence times is the fact that many of the divergence times are within the same order of magnitude. Many unsatisfactory estimates have been recognized and rejected. It is like walking into an old-fashioned horologist's shop with no idea of the time, and through the cacophony of ticking and murmuring of electric pulses seeing that some of the clocks are clearly dancing to their own music of time, while a majority are reading some time about half past two. You could not be certain of the exact time, and certainly not that it was half past two, but you could feel pretty sure that it was not before one o'clock, nor yet teatime. So with the molecular "clocks." It seems likely that the far distant common ancestor of the line that led to starfish and man on the one hand, and trilobite and fly on the other,* lived at some time between 750 million and 1250 million years ago—after lunch in the history of life, but before teatime. This common ancestor probably carried a pair of primitive eyes.

If this figure is even approximately correct, it places trilobites more than 250 million years after the origin of eyes, and quite possibly at about 500 million years. Trilobites offer visible evidence of the halfway stage in eye development, a testimony to the continuity of the genes that still intervene in the development of every embryo. Inspired by our modern genetic knowledge, we can feel a bond with the trilobite that would not have been apparent when the nineteenth-century investigators first looked into those stony eyes. To these earlier scientists, the trilobite was an alien creature, whose connection to the world of living animals was remote, almost imponderable. They might have perceived some thread of common ancestry, but I doubt that they intuited that aspects of the design of a trilobite were perpetuated in our own

---

*This is correctly known as the join between the Protostome and Deuterostome animals. The former includes all the arthropods, as well as molluscs and many worms, while the latter comprises the vertebrates (including us) and echinoderms (sea urchins and relatives). The distinction between these great assemblages of animals was recognized by nineteenth-century embryologists as a fundamental difference in body development; it has stood the test of a century of biological investigation and, more recently, molecular analysis. Recently, the Protostome idea has been refined by the recognition of a large group of animals that have moulting hormone.

embryos. An advance in knowledge has served to entwine us closer with the past. "Look into my eyes," the trilobite seems to say, "and you will see the vestiges of your own history."

This shared history of vision is far from trivial. In our visually-dominated world sight is almost synonymous with understanding. We acknowledge light dawning by saying: "I see!" The metaphor of vision suffuses our attempts to convey comprehension: we bring issues into focus, we clarify our views, we sight our objectives, we *look into* things. We accept the evidence of our own eyes. The conjurer turns the veracity of sight head over heels: now you see it—now you don't. We find his tricks disturbing because we are so wedded to the truth of sight. To understand the deep history of vision gives us a way of perceiving how even the most distantly-related animals, far off in remote geological time, managed to comprehend their world. We can describe how they understood their vanished seascapes in terms that we can still apply to our own habitats, a compound of sight, images and colour. For us to see that the trilobites once saw, too, is to bring them within the compass of our own understanding.

Trilobite eyes are made of calcite. This makes them unique in the animal kingdom.

Calcite is one of the most abundant minerals. The white cliffs of Dover are calcite; the bluffs along the Mississippi river are largely calcite; the mountains stacked like giant termite mounds in Guilin Province, China, are composed of calcite that has resisted millennia of weathering. Limestones (which are calcite) have been used to build many of the most monumental and enduring buildings: the sublime crescents of Bath, the pyramids of Gizeh, the amphitheatres and Corinthian columns of classical times. Polished slabs composed of calcite deck the floors of Renaissance churches in Italy, still grace the interiors of Hyatt-Regency hotels, or conference halls, or wherever architects wish to suggest the dignity that only real rock seems to confer. Rubbly limestone builds our rockeries; its finer, whiter counterpart provides the raw material from which great sculpture grows. Only silica sand seems as ubiquitous. Surely one could expect no surprises from a substance so common and so familiar. Yet it was calcite transformed that allowed the trilobite to see.

The purest forms of calcite are transparent. In building stones and decorative slabs it is the impurities and fine crystal masses that provide the colour and design: the yellows and greys and fine mottling. The dark red of the *scaglio rosso* so typical of Italian church floors is a deep stain of ferric iron. Purge calcite of all these impurities and it is colourless. But it may not be transparent even then. Chalk is almost pure calcite, but it is a mass of tiny grains—fossil fragments most of them—which scatter and reflect the light: hence its almost indecent whiteness. When the Seven Sisters on the southern English coast emerge from a sea mist it is like observing a line of undulating starched sheets, so frigid is their purity. But when a calcite crystal grows more slowly in nature, then it may acquire its perfect crystal form, and be glassy clear. The chemical composition, calcium carbonate ($CaCO_3$), is simple as minerals go. As the crystal grows the constituent atoms stack together in a lopsided way, and do not allow other stray atoms to intrude to cloud its mineral exactitude. Layer builds on layer to reveal the crystalline form, the macrocosm of the gem reflecting exactly the microcosm of atomic sructure. As with the handiwork of a master mason, there is no mistake permissible in the atomic brickwork. Large, fine crystals often grow in mineral veins. These are often rejected by miners in search of rarer booty, for precious metals sometimes hide in grey and opaque minerals that seem dull by comparison with calcite's perfectly formed spar. Some of these crystals are sharply pointed and then are described as dog's tooth spar, looking much like the zig-zag ornament favoured by Norman craftsmen over church doors; others, blunt-tipped, are termed nail head spar. But the clearest crystal, transparent as a toddler's motives, is Iceland spar.

Look into a crystal of Iceland spar and you can see the secret of the trilobite's vision. For trilobites used clear calcite crystals to make lenses in their eyes; in this they were unique. Other arthropods have mostly developed "soft" eyes, the lenses made of cuticle similar to that constructing the rest of the body. Within this limitation there is enormous variety: many-lensed eyes like those of the fly; large complex eyes such as those of most spiders; eyes that can see in the dark; eyes that function best in brilliant sunshine. The octopus among the molluscs has an eye that is famously like that of backbone-bearing animals, and provides one of the best examples of convergent evolution in the animal kingdom. Most of us will have contemplated the sorry eyes of a dead fish, and remarked the comparison with our own, large, focusing eyes. Trilobites alone have used the transparency of calcite as a means of transmitting light. The trilobite eye is in continuity with the rest of its shelly armour. It sits on top of the cheek of the animal, an *en suite* eyeglass, tough as clamshell.

The science of the eye demands a little explanation. It all depends on the optical properties of calcite, and this depends in turn on its crystallography. If you break a large piece of crystalline calcite it will fracture in a fashion related to its fine atomic structure: such cleavage of the mineral does the bidding of the invisible arrangement of matter itself. You are left with a regular, six-sided chunk of the mineral in your hand, termed a rhomb. Neither foursquare like a cube, nor rectangular like a chunk of chocolate, the sides of a rhomb lean away from the perpendicular. The geometry of mineral shape can be described quite simply by the orientation of a few main axes passing through the centre of the crystal: the simplest case is the cube, in which axes passing through the centre of the faces and meeting at the middle are all at right angles and all the same length. These axes are termed a, b, and c, respectively, a case of science for once taking the simplest route to make a name. In the structure of calcite, one major axis has three axes perpendicular to it set at 120 degrees from one another, hence the configuration of the rhomb. The clear calcite of this not-quite-a-cube treats light in a peculiar way. If a beam of light is shone at the sides of the rhomb it splits in two; this is known as double refraction. The rays of light so produced are the "ordinary" and the "extraordinary" rays: their course is determined, just like the shape of the rhomb, by the stacking of the individual atoms. There is a huge specimen of Iceland spar on the first floor of London's Natural History Museum through which you may peer to see two images of a Maltese cross, one generated by the extraordinary, and the other by the ordinary rays. But there is one direction, and one direction only, in which light is not subjected to this optical splitting. This is where a ray of light approaches along the *c* crystallographic axis; from this direction it does not split into two rays at all but passes straight through.

The way that calcite treats light might have remained no more than an odd fact to be trotted out as an esoteric answer in tests of general knowledge. But what the selectivity of the *c* axis guarantees is that light approaching from the angle at which it points is specially favoured. If a crystal is elongated in parallel to the *c* axis into the shape of a prism light will still pass unrefracted through the crystal along the long axis of the prism. But light approaching the same prism from other angles will be split into ordinary and extraordinary rays, which will in turn be deflected to reach the edge of the prism, where they might be partly internally reflected, or refracted yet again. When the prism is long enough the *only* light to pass clearly through to the far side of the prism is that which approaches from the direction of the *c* crystallographic axis. To put it the other way round, the light that such a crystal "sees" approaches from one particular direction. It is an astonishing fact that trilobites have hijacked the special properties of calcite for their own ends. They have crystal eyes.

The eyes of trilobites are composed of elongate prisms of clear calcite. Most eyes have many such prisms stacked side by side. By comparison with dozens of other kinds of arthropods the prisms obviously functioned as individual lenses, in just the same way as a fly's eye is a honeycomb of hexagons, each one a lens—or the dragonfly's, or the lobster's. The trilobite carries on its head another example of such an arthropod compound eye—an eye composed of numerous small ocular units, which had to collaborate to paint a portrait of the world. Each component unit is a lens. The unique difference is that the trilobite's lenses are composed of a rock-forming mineral. It would be no less than the truth to say that the trilobite could give you a stony stare. One is reminded of the strange lines from that strangest of Shakespeare's plays, *The Tempest*:

> *Full fathom five thy father lies:*
> *Of his bones are coral made:*
> *Those are pearls that were his eyes:*
> *Nothing of him that doth fade,*
> *But doth suffer a sea-change*
> *Into something rich and strange.*

If to travel back to the time of the trilobite is a historical sea-change then there can be nothing stranger than the calcareous eyes of the trilobite. And pearls are chemically the same as the trilobite's unblinking lenses, being yet another manifestation of calcium carbonate, although pearls are exquisite reflectors of light rather than transmitters of it. The weirdness of Shakespeare's line results from his suggestions of pearly opacity, the hints of a corpse transformed; dead, yet seeing. The trilobite saw the submarine world with eyes tessellated into a mosaic of calcified lenses; unlike the dead seafarer, his stony eyes read the world through the medium of the living rock.

Trilobite lenses were orientated so that the c crystallographic axes ran along the length of the prisms comprising each lens. In most lenses this axis is exactly at right angles to the surface of the lens. If you can see the whole surface of an individual lens (perhaps using your own hand lens) then the chances are that the lens could look back at you. The lens, of course, can not itself see. But it permits light from a favoured direction to pass through it. The average trilobite eye was a stacked mass of tiny, elongate prisms, each pointing in a subtly different direction. A long, semicircular eye might have hundreds or even thousands of such lenses. Some of the lenses might have their *c* axes pointing forwards; some sideways; some backwards. One has to imagine all the *c* axes poking out of the centre of the lenses like a battery of tiny needles. A large eye would carry a veritable hedgehog (or porcupine) of such imaginary needles: each needle can be thought of as representing the beam of light that could pass through the lens, like a host of tiny arrows each with a particular target. Every arrow of light contributed a small ray of understanding to the eye, each lens having its own dedicated field of view.

It is very likely that the trilobite eye functioned in the same way as that of living arthropods with compound eyes. So we might expect to find at the base of each lens a receptor cell which could respond to an incoming ray of light. These cells were evanescent, as fragile as the rocky lenses that lay above

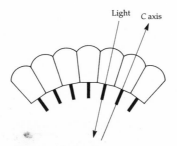

How the trilobite's eye works. Rays of light passed through calcite lenses in the preferred direction parallel to the major *c* crystallographic axis. The light receptors lay on the inside of the eye.

them are permanent. They cannot be preserved as fossils, but their existence was surely necessary to turn a mechanical agglomeration of beams of light into images. Light of itself cannot generate understanding, any more than an image reflected in a pool interprets the scene it so faithfully duplicates. Information has to be gathered by nerves, and interpreted by brains. Because of the selective field of view from each lens the ancient world of the trilobite must have been perceived as a mosaic, a shuttle of tiny images, overlapping, subtly changed from lens to lens. The resolution of the image must have depended to some extent on the number of lenses. A more detailed perception was possible with many lenses—the more, the better. Hence it is not particularly surprising to find that some trilobites had an almost unreckonable number of tiny lenses.

One of the most difficult jobs I ever attempted was to count the number of lenses in a large trilobite eye. I took several photographs of the eye from different angles and then made enormous prints magnified large enough to see individual lenses. I started counting as one might: "one, two, three, four" . . . and so on to a hundred or two. The trouble was that you had only to look away for an instant, or sneeze, to forget exactly where you were, so it was back again to "one, two, three . . ." Teeth were gnashed, imprecations muttered, deities' names taken in vain. Eventually I hit upon the notion of pricking each counted lens on the photograph with a pin, so that it wasn't counted twice. The trouble was moving to the next photograph: what was the last lens that I'd identified and how did they link from one picture to the other? Was it that one with the little scratch, or that one a mite larger than its neighbour? The work was undeniably suitable for an obsessive with insomnia. I got to a total of more than three thousand before I vowed that, in future, I would simply estimate the number of lenses in a bit of an eye, and use my best arithmetic to estimate the whole number.

Across their many species the number of lenses in trilobite

eyes varies from a mere one to several thousands. No doubt the effectiveness of the eyes varied accordingly. But, large or small, they received light along the *c* axes of the calcite crystal of which they were made.

This fact carries an interesting corollary. If we know where the light came from to pass through an individual lens then we also know exactly where the trilobite could look. Turn the tiny arrows of the *c* axes around and they dart out into waters of the marine world surrounding the trilobite, spearing any object in vision. To understand what the animal knew of its environment we only have to sum the lines of vision of its lenses. We can look through the eyes of the trilobite to see the world as it was observed hundreds of millions of years ago. Eyes made of pure crystals are attuned to the images of ancient scenes. Lenses ranked in horizontal lines may just see a horizon, whereas curved eyes with many lenses may see a wider view. Learn where the lenses face and you will have a prospect of a trilobite's field of vision.

The first investigator of details of the field of view of trilobite eyes was Euan Clarkson from the University of Edinburgh. Euan always refers to trilobite eyes as "peepers" after the line in the song:

> *Jeepers, creepers!*
> *Where'd ya get those peepers?*

What he did was to mount trilobites in such a way that he could with great accuracy measure the direction of the pole (*c* axis) of each lens. Then he plotted the spread of directions of these axes on to a stereographic net, which is a way of displaying which part of the full 360° sphere of vision the trilobite actually utilized. He saw what the trilobite saw.

The majority of the trilobites that Euan studied did not see all around them. He proved that many common trilobites preferred to see alongside. The eyes looked sideways and forwards and often a little backwards: they glanced askance. But why? The field of view was directed over the area surrounding the animal, as it might be a searchlight sweeping over the ground and low bushes, but unconcerned with the province of the sky. There is a simple explanation of why their field of view was limited in this way. Most trilobites lived on and around the sea-floor, and this was the world they wished to appraise. It was a world where enemies approached on scuttling limbs across the surface on which they lived, where potential food may have lain half buried in soft sediment, or slowly crawled or ambled over the surface of the mud. A likely mate was a possible neighbour working an adjacent plot on the sea-floor, but deserved a closer inspection, just in case. A rival might come sidling up at any minute, and needed to be spotted before acquiring the advantage of surprise. Antennae swept the water in front of the advancing animal to sniff or taste any chemical signals born on currents which might complement the evidence of their own eyes; tactile and olfactory senses played their age-old complementary role to visual acuity. It was a world atop the sediment, where most of the events of the day or night carried on in the same small hemisphere.

The modern equivalent of this world is still present everywhere on muddy sea-floors—but does not have the sort of glamour to rate the television coverage of coral reefs. It is a place where a variety of humble worms process nutrients from the sediment itself, many more burrow within it, still others stir it up to make a nutritious soup. It is an environment where innocuous grazers are hunted by sneak thieves and footpads, where some animals disguise themselves as seaweeds, still others breed so fast as to outstrip their hunters. It is a world full of the subterfuge of survival, all fed by the organic richness of the sediment itself; a sideways world, where you watch out for your neighbour, for he may not be what he seems. It is no wonder that the average trilobite was concerned with its muddy environs: whether it was hunter or hunted it had to look out across the undulating view, twitching and perceptive, for its life depended upon it. For most trilobites their eyes were a key factor in their survival (although we have met some blind ones). Who can fail to be impressed by the trilobite's petrified tokens of the triumph of vision one hundred and fifty million years before even the first tentative excursions had been taken by plants towards colonizing the land?

Look closely at a trilobite eye and you will see a honey-comb of tiny lenses. Like many packed things in nature, the lenses are mostly hexagonal. They follow the bidding of geometry just as many corals do, or insect eyes, or even many patchwork quilts. Where small, similar objects squeeze to-gether cheek by jowl until they touch and jostle for accommo-dation they naturally tend to compress into hexagons. It is a way of equalizing the pressure for space with all the neigh-bours. The centres of adjacent hexagons are all equidistant from one another. So the average trilobite lens is long and thin, a few tens of thousandths of a millimetre (μ) across, with its c crystallographic axis running along its length, and is hexagonal. If the eye were perfectly flat, its design would be as uneventful as that of a sheet of patterned linoleum. But the geometry of "bending" a sheet of hexagons around a curved surface is not straightforward, and you will find the occa-sional odd-shaped lens or a shuffle in the lines of lenses, to make a little space around a curve. (We all know the compro-mises that have to be made when wrapping a football in Christmas paper.) But even so, some trilobite eyes seem to be astonishingly regular, lines of hexagons making gentle spiral curves that run obliquely across the eye from bottom to top.

Euan noticed something else about the construction of the trilobite's eyes: the smaller lenses were concentrated at the top of many eyes. The eye surface—known as the corneal sur-face—had to be moulted along with the rest of the animal's hard exoskeleton as it grew. The eye itself grew in size in har-mony with the rest of the animal: more lenses were added after each moult as the new skeleton hardened. New crystals were added in from the top of the eye in a zone of generation. With successive moults these lenses were incorporated into the main body of the eye, passed downwards in a graded chain. These differences in lens size also helped to maintain the regularity of the design across the curved surface of the eye. It is fiendishly clever (as Hercule Poirot would to say) that these "primitive" animals could play such games with the mineral world in the service of eye geometry.

We cannot know *exactly* how the trilobite saw with its crystal eyes: the nerves have left no trace. It is like finding an artefact from some remote civilization—we may be able to assume its general function, but we will never know the thoughts that passed through the mind of its user. The trilobite will always keep a certain distance from us; there will be limits on the intimacy we can attain. What we can guess is that the honeycomb-like trilobite eye may have permitted comprehension of the world in the same fashion as the similar compound eyes of living arthropods. Apposition eyes do not form complete images of their surroundings (some other arthropod eyes have lenses arranged in such a way that they are able to collaborate and produce a single, complex image). Densely-lensed eyes of trilobite type are particularly good at detecting movement. Another animal approaching across the sediment surface will trigger one lens after another as its image impinges on different parts of the field of view. If the change is alarming the trilobite may be stimulated to take evasive action: maybe to roll up into a ball or to swim away as fast as possible. To look through the eye of the trilobite is to see the world as fragmented pieces of information—trilo-bytes, I am tempted to dub them. The animal was not able to see as we see, but appreciated the world in a thousand fragments of light, as if the brain were a *pointilliste* with a palette of prisms.

There is still more to tell of crystal eyes. Although the great majority of trilobites have eyes such as I have just described, there are some which are obviously different. One of the commonest trilobites in the Devonian rocks of New York, Ohio and Ontario, and also in Germany and Morocco, is the compact beast called *Phacops*, which we have already met in the parade of trilobites in the last chapter. Moroccan *Phacops* can be bought for a handful of dollars and they are cheap at the price. If you work in a large museum *Phacops* is one of the most frequent trilobites to walk in through the door accompanied by its owner. One greets it as one might an old friend. It is always a pleasure to point out its large crescentic eyes, that stand proud of the cheeks like the retractable headlamps that grace the front of a Porsche. But wait! There is something peculiar about these eyes. Instead of the lenses being so minute that it requires a microscope to see them properly you can discern them quite clearly unaided. To the naked eye they look like a series of tiny, perfectly formed balls—a real case of *those are pearls that were his eyes*. The lenses line up quite conspicuously in vertical files, often with a little space between them, and they are stacked in the manner of hexagons, so that any lens has six neighbours. It is another example of close packing, and not different in principle from that displayed by lenses of other eyes. But it is the regularity of the eye that is so striking. We are accustomed to expect a little sloppiness in nature's designs: the leopard's spots are hardly mechanically repetitive, no two snakes have identical zig-zags on their backs. But these eyes seem to have been turned out by a machine, neat as billiard balls arranged in a box. They are obviously something different from the minutely lentiferous eyes of the usual trilobite. Instead of averaging many hundreds to thousands, this kind of eye had lenses which numbered a hundred or so, or could even be counted on the collective fingers of the average family.

If the trilobite eye is something out of the ordinary, then the phacopid eye is odder still.* One way to study it more closely is to cut a section through the lenses, so that the optical

---

*The "normal" kind of trilobite eye is known in the trade as holochroal; the special kind that *Phacops* and its relatives had is termed schizochroal.

properties can be studied under a high-powered microscope. Even though the animal has been dead for so long, there is a hint of the sacrilegious in taking one of these beautiful creatures and cutting across its head with a circular saw. Those serried pearls which have lasted inviolate for several hundreds of millions of years may now be destroyed in an afternoon.

But sections made in this way reveal strange secrets. First, the lenses are indeed nearly spherical, or perhaps slightly drop-shaped. A phacopid lens has a disquieting resemblance to a glass eye. As a student I once did a labouring job alongside an old boy with a glass eye, who would pop it out whenever there was a lull in the conversation, toy with it for a bit, and then pop it back in again. In or out, it made no difference—he could not see through it—whereas the schizochroal lens was evidently functional. Nor could it be dislodged and replaced, as it was sealed in a solid sheet of calcite (although, like everything else on the trilobite, it would have been moulted). Second, there is usually a little "wall" between adjacent lenses, a kind of baffle which stopped light from one lens overlapping with that of the next. Often, the lenses are slightly sunken, and the areas between the lens are a little swollen. The optical arrangement is clearly a very sophisticated structure which quite belies the antiquity of the animal. This may come as something of a surprise: we might expect an eye from half-way along optical history to have a slightly slung-together look, or at least broadly to resemble the eyes of many other lowly animals, as does the run-of-the-mill trilobite eye. But the eye of *Phacops* is something unexpected, a sports coupé in the age of the boneshaker. Not only does it have calcite lenses, but they are of a singular type.

Surely such specialized eyes must have functioned in a very specific fashion. In the living fauna there are no really convincing analogies: one investigator drew attention to the ant-lion larva which has somewhat similar drop-like eyes—but not made of calcite crystals. It was an American investiga-

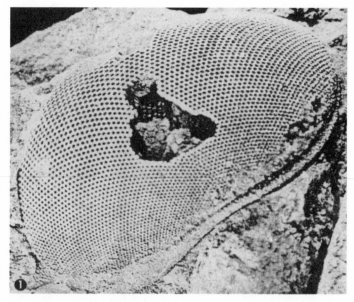

Trilobite eyes. The holochroal eye consisting of numerous hexagonal lenses (above)—*Pricyclopyge,* an eye adapted to detect the smallest movement; (below) the specialised schizochroal eye of *Phacops* showing fewer spherical lenses, each one finely tuned to its habitat. (Photographs courtesy Euan Clarkson.)

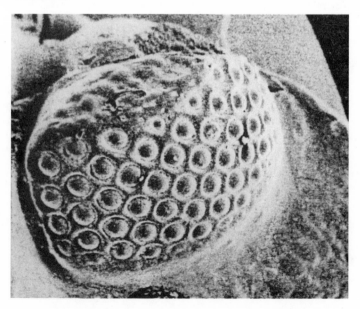

tor at the Smithsonian Institution in Washington, Kenneth M. Towe, who in 1972 demonstrated the efficiency of the phacopid eye lenses in the most graphic way: he took photographs through them.

When you are a scientific visitor to the Smithsonian Institution, the National Museum of Natural History, you enter with the crowds through the public entrance, but then wheel off to one side to telephone through to your contact behind the scenes. Within a few minutes you pass through an inconspicuous door into a different world of cabinets and collections, a cool, scholarly enclave away from the metropolitan throng. When Ken Towe worked there his office had a view across the grand avenue to the FBI building. Visitors to the Smithsonian were taken there for lunch, an experience so unexciting as to defuse any paranoia about spooks and fifth columnists. Using the trilobite lens as a substitute camera lens Ken photographed the FBI building—not perfectly, but recognizably. What more curious tribute to J. Edgar Hoover than to have his workplace photographed through the eyes of an ancient fossil! Another photographic attempt successfully captured the grinning "happy buttons" that were in vogue at the time. It was as transparent as calcite that the phacopid lens could form sharp images, bringing into focus objects of different sizes at varying distances. The phacopid lens saw larger pieces of the world than the tiny lenses of most trilobites, and saw them clearly. It was an astonishing feat of optical engineering accomplished using calcite, the most quotidian of minerals.

How this trick might have been done was discovered by Euan Clarkson and Riccardo Levi-Setti not long afterwards. It was already apparent from the spherical structure of the lenses of *Phacops*—and their large size—that some different method was being used by these trilobites to form their images when compared with the tiny lenses of their relatives. These were fat, biconvex lenses, designed to bring beams to a focus. If you hold a clear glass marble up to the light and peer through it you can get some idea of the process: you will see an upsidedown world, all bent and distorted. The trilobite images seem to be much clearer than that. How could this be? The problem with light travelling through a convex lens to a focus is that different rays travel different distances through the lens according to their trajectory—and in a refractive material like calcite that means that they get bent to different degrees: the result is a fuzzy focus. Like my old workmate's glass eye, to be transparent is not enough to see. The technical term for this fault in design is spherical aberration.

Riccardo Levi-Setti is a nuclear physicist at the University of Chicago, a place where *everyone* is almost effortlessly brilliant. He also has a private passion for trilobites, which he pursues more vigorously than many a professional palaeontologist. Euan and Riccardo made an interesting combination: an extravagantly hairy and amiable Scotsman and a well-groomed and suave Italian. What they discovered together was that *Phacops* had solved the problem of spherical aberration. Euan had made out a kind of bowl within the individual schizochroal trilobite lens and at the base of it, and he identified it as part of the lens with a different structure. In some kinds of preservation this bowl would weather out separately, so that the eye came to resemble a series of little dishes. Euan and Riccardo discovered by making thin sections that in this part of the eye something strange had happened to the calcite: it had become impure. Some of the atoms of calcium within the crystal structure had been replaced by atoms of its closest elemental relative—magnesium. Because the atoms are so similar, magnesium could sneak in, like a spy in an appropriate uniform infiltrating an army. Even in the purest calcite there are a few such hidden agents. The effect of this process continuing far enough to make "high magnesian calcite" is to alter the refractive index, the capacity of the crystal to bend light. With a wonderful fineness of balance the thickness of the high magnesian layer varied across the lens in just the right degree to correct the spherical aberration—for every bend to the left a compensating bend to the right. This corrective layer was what made the bowl. The trilobite had manufactured what a modern optician might term a doublet, wherein two wrongs do make a right because of the way they are spliced together.

As a grace-note on this discovery, Riccardo noticed that the trilobite's design had been anticipated by the great seventeenth-century Dutch scientist Christian Huygens (1629–95) and the French polymath René Descartes (1586–1650). They had sketched out an optical "cure" for spherical aberration in a lens which proposed a compensating bowl designed almost exactly like that of the trilobite. This may indeed be a wonderful example of Art imitating Nature, or perhaps rather of Nature anticipating Science—by more than 400 million years. S. J. Gould commented in an article in *Natural History* in 1984 that "the eyes of trilobites . . . have never been exceeded for complexity and acuity by later arthropods . . . I regard the failure to find a clear 'vector of progress' in life's history as the most puzzling fact of the fossil record." The point Gould makes is that it is hard to see how the trilobite could have achieved its optical design in a still more sophisticated fashion; there remains a feeling that arthropods ought to have learned some cleverer visual tricks since the Devonian. The notion of progress in life's story is an intellectual quagmire. It embodies a belief in "improvement" which is difficult to defend. Maybe the trilobite should stand condemned for having perfect eyes while its limbs were decidedly second rate. Or maybe we should berate it for carrying such a heavy suit of armour, while at the same time acknowledging its nonpareil peepers. If you put yourself in the right frame of mind you can imagine the trilobite like one of those medieval knights, cumbersome and unwieldy, however well-protected. We might be able to talk ourselves into imagining a story of progress whereby sleeker warriors outclass the lumbering, articulated Sir Phacops. Serve him right! Progress is the thing!

This is all nonsense, of course. The eye of these phacopid

Crystal Eyes

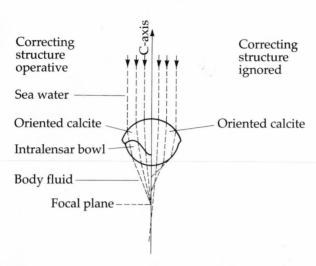

The drawing made by Euan Clarkson and Riccardo Levi-Setti to illustrate how the more highly refractive bowl inside the lens of *Phacops* helps to bring rays more closely to focus.

trilobites is wonderful, and unique, but I do not know how to measure it against the lustrous eye of the dragonfly, that can so resolve an image as to snatch a wasp on the wing. I cannot say that it is better or worse than the eye of those marine crustaceans who use silvered boxes to focus weak light at depth into a precise image. I do not know how it stands in comparison with the several, surprising eyes of spiders. Who is to calibrate progress, who to legislate on the unit of improvement? The trilobite was no doubt a perfect citizen of its times, and its eyes focused on the problems of its daily life sufficiently well to allow the beasts to throng in their thousands on the sea-floor. The astonishing fact is not that the eyes were so perfectly engineered but that the sea rewarded such a specification even then. We cannot name a time when the trilobites reached their acme, after which there was either stasis or decline. Life just isn't like that.

My own engagement with trilobite eyes was an investiga-

tion of goggle-eyed species. One of the most peculiar specimens I had discovered in the Ordovician rocks of Spitsbergen
was quite unlike other trilobites that I had seen illustrated in
textbooks. It was long and thin, and the axis occupied much of
the body, so that the pleurae were reduced to little triangles.
But the eyes were truly extraordinary: they were enormous
and inflated, puffed up like little bladders. It was weeks
before I was able to convince myself that I had located the correct free cheek for this trilobite (I had no entire specimens, so it
was the usual matter of completing the jigsaw puzzle). But in
the end there was no question: the eye had become so enormous that it abutted against the edge of the cranidium in an
almost straight line running alongside its entire length. Only
one eye among the selection I had cracked out from the
Ordovician limestones had the right "fit." I was now able to
place the eyes back in position on the headshield to reconstruct the skeleton of the entire animal. Now it seemed even
odder. It was obvious that the eyes bulged out on either side
of the head in the manner of those slightly grotesque orna-
mental goldfish that have such a thyroidal look; in proportion
these "peepers" were even larger—virtually the whole free
cheek had turned into an enormous eye! What could be going
on? I named this strange animal *Opipeuter inconnivus*, having
recruited the help of a classicist friend to find out the Greek
for "one who gazes"—*Opipeuter*; *inconnivus* means "without
sleeping." Trilobites, of course, were unable to blink.

There were some other details that attracted my attention.
When the eyes were "fitted on" it was perfectly clear that they
hung down well below the level of the rest of the animal. If
you view the majority of trilobites from the side the base of
the animal makes a line parallel with the sea-floor on which
they lived; not so *Opipeuter*. Furthermore, the edges of the
cheeks were sharp, with their cutting edges directed downwards. This is where Euan's work on the direction in which
crystal lenses could see became so useful. The lenses on
Opipeuter were tiny and of the crowded, hexagonal type used

by most trilobites, not the special *Phacops* type. Hundreds upon hundreds of lenses crowded together over the bulging surface of the eyes. But they differed from the crescent-shaped eyes of nearly all the trilobites I had ever seen, that looked predominantly sideways across the sea-bed. In *Opipeuter* there were lenses that faced sideways, of course, but there were also many dozens of lenses that "looked" forwards. If I was right about the bulging orientation of the eye, there were nearly as many lenses that were capable of looking upwards as well—and even downwards. And with the thorax so trimmed at its edges it was likely that the eyes projected out far enough to command a view backwards... there were lenses gawping every which way. These weren't just peepers, they were more like oglers.

Surely this trilobite needed to see all around, but what could it possibly be for? Where in the ocean is it necessary to have an all-encompassing view of the watery world? Perhaps it was my customary view of trilobites as bottom dwellers that prevented me from seeing the obvious. It must, of course, have been a swimmer! A leap of the imagination had the trilobite leaping off the sea-floor. *Opipeuter* had the freedom of the Ordovician oceans: it needed to see everything. Suddenly there was a different vision of the lives of trilobites: from grovelling on the sea-bed, they filled the seas as well. The former oceans could have swarmed with trilobites, just as krill throng in living seas. That was why the body of *Opipeuter* was long and thin compared with most trilobites, and why its design was so poor for resting on the sea-floor. It had a vaulted axis to house the muscles used to power its swimming limbs, but economized on the rest of its shell so as not to overburden the work of the sculling appendages. Some of the rocks in Spitsbergen were almost made of this trilobite and its relative, *Carolinites*, so that it was not hard to imagine a sea alive with thousands of these little animals, swimming in the brilliant sunlight while, far below on the sea-bed, *Triarthrus* slowly ambled through the soft mud.

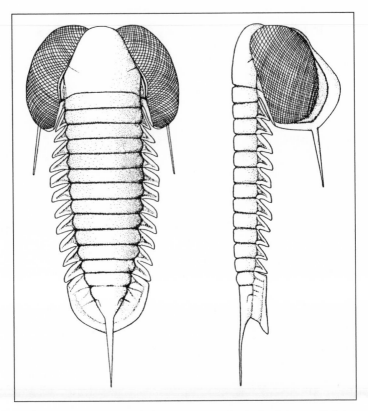

My reconstruction of the giant-eyed Ordovician trilobite *Opipeuter*—swimmer across the ancient oceans—from the top and the side.

There proved to be a number of different trilobites with this free-swimming design; the one-eyed *Cyclopyge* had been noticed already as a swimmer by the great early twentieth-century geologist Eduard Suess. He made a comparison with some huge-eyed, living crustaceans. When I discovered this observation, buried in the middle of his great work *The Face of the Earth*, I understood that ideas are seldom ever really new. *Cyclopyge* was a more compact animal than *Opipeuter*, with fewer thoracic segments, but it lacked nothing in the eye department. It, too, was found in considerable numbers in company with several other genera of bug-eyed trilobites. I have studied these cyclopygid trilobites in Wales and Bohemia. They were collected from dark mudstones that were originally deposited in relatively deep water, to which they are apparently confined. I have sat under hedgebanks with rain dripping down my neck breaking open dark shales near Carmarthen—in Welsh, "Merlin's Castle"—and could not have been more amazed if the old magician himself had popped out of the undergrowth than by my first sight of the fat eyes of *Pricyclopyge* (p. 105), staring out at me after an incarceration of 470 million years, some of its lenses still glistening slightly. In contrast, many of the limestones which yielded *Opipeuter* and *Carolinites* were laid down in shallow seas—there were other fossils with them that proved it. Could it be that *Cyclopyge* and its friends swam at depth in a world of pervasive dimness, while *Opipeuter* and its allies lived in the surface waters of the sea in brilliant light?

I was able to test this idea thanks to a grant of money from the Leverhulme Trust. It is fortunate for me that a few charitable bodies still exist who will support what has been called "blue skies" research—that is, research that has no industrial or commercial spin-off. I doubt you can get bluer skies than trying to find out about the optics of Ordovician oceanic trilobites. Leverhulme supported a young post-doctoral assistant, Tim McCormick, to immerse himself in trilobite eyes for several years. We had discovered that it was possible to deduce the light intensity in which an arthropod customarily lived by making some very careful measurements on the compound eyes. This had been worked out on a variety of living species—and it seemed a good idea to try it out on trilobites. Tim had to mount perfectly preserved specimens so that he could measure such details as the distance and angle between adjacent lenses, so as to obtain a mathematical quantity termed the eye parameter. It was a labour that lasted for six months. Imagine our delight when our alleged surface swimmers came out with the right value for bright illumination, and the trilobitic Cyclops condemned itself to live in gloom. Thus we were able to insinuate ourselves into the daily lives of our favourite animals. We had no Steven Spielberg to animate these creatures into plausible simulacra. Our vision of trilobite lives remains lodged in our own imaginations, but all the more intense for that. Now we can visualize an Ordovician sea with a clarity that would have astonished Adam Sedgwick, James Hall or Sir Roderick Murchison. We can see—literally, through the eyes of the trilobite—an ocean teeming with swimming animals, some grazing on tiny zooplankton near the surface, others deep down in a twilit world, and both above a sea-floor where a profusion of other trilobites again scurried or ambled.

Nor was this the end of our vision. Among the small, deepwater cyclopygids, there were rare examples of a different and larger kind of animal. This trilobite, too, had big eyes, but they did not bulge out; instead they were tucked into the side of the head. And what a peculiar cephalon! It was extraordinarily long, largely because it was produced forward into what can only be described as a snout or nose. This "nose" comprised the front part of the glabella (which was itself unusually smoothed out, and had lost all its furrows) and an extended section of doublure beneath it; together they made something that looks like the "nose" of a dogfish or small shark. The rather large pygidium had a dished appearance. The whole thing was beautifully smoothed out, like a torpedo. The shape of the animal reminded me of an illustration I had once seen in a textbook of an idealized hydrofoil—a shape designed to cut through the water with a minimum of frictional resistance. Could this trilobite have been the porpoise of its kind? I needed advice.

Just up Exhibition Road from the Natural History Museum is the famous Imperial College of Science and Technology. Imperial has long been one of the leading institutions for all the sciences that make things work: somewhere there, there must be a B & Q Professor of Gadgetry. I soon learned of a lecturer called David Hardwick who could help me design an experiment to test the streamlining of my dogfish-nosed trilobite (which goes by the inelegant name of *Parabarrandia*). In the hydraulics department there were all manner of flume tanks and watery experimental gear. The simplest thing that could be done was to suspend life-sized models of various trilobites in a transparent-sided tank, along which a current flowed. If you then streamed a dye into the tank you could see which animal shapes produced the best streaming past the body. Most of the models we tried had all kinds of turbulent wakes: projecting eyes, for example, showed little coloured eddies behind them. Now we could see why it was an advantage for *Parabarrandia* to have its eyes flush with the sides of its body. In fact, the dye streams flowed past this trilobite as if they were long, straight tresses swept out in a breeze: it was a wonderful demonstration of the streamlining theory. We could now imagine this animal elegantly outpacing all other swimmers in the Ordovician. But we needed proof to convince our sceptical colleagues. So we now went further by designing a way of measuring the drag itself, by seeing how far carefully scaled models were deflected when suspended in a gentle current—the poorly streamlined species would experience the greater deflection as their bodies offered resistance to the movement of the current. This required an experiment in an open-topped tank. The trilobites were suspended in the tank as if they had been bait for some Palaeozoic fisherman. Now the current was turned on, and the deflection measured with a travelling microscope. In a white lab coat and with my sensitive measuring equipment I felt like a real boffin.

I nipped out for a quick cigarette (I was a smoker then) while the experiment was still running and was horrified when I returned to find the whole laboratory floor several feet deep in water. I had visions of being banned from Imperial College for ever, and having my screwdriver confiscated by

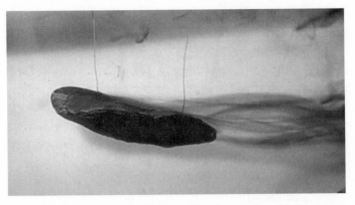

The free-swimming trilobite *Parabarrandia* (Ordovician, Czechoslovakia) immersed in a flume tank. The stream of dye to the right shows the excellent streamlining.

the B & Q Professor. Mortified, I sought the aid of the laboratory technician. Giving me that look which I have often received from garage mechanics, he waded into the middle of the floor and pulled out the most enormous bath plug I have ever seen. Within minutes the water had gurgled away down the plughole, while I stood by with my mouth hanging open slightly, damply rueful. But the experiment had proved the point. There were streamlined trilobites that sped through the Ordovician ocean.

It is a curious fact that, despite the beauty and complexity of trilobite eyes, it seems curious that plenty of trilobites could do perfectly well without them. A large number of them were blind. Eyes, apparently, could be readily jettisoned. In some examples we can even show how it happened. An ancestral species had large eyes, and a succession of daughter species had smaller and smaller ones, until the facial sutures ran across the cheeks leaving no sign of lenses at all. Some of the trilobites related to *Trinucleus* had just one lens left perched high on the cheek. My colleague Bob Owens from the National Museum of Wales and I collected in some localities in South Wales where ten or more blind, or nearly blind, trilobite species lived together, and they must have crawled about the sea-floor in a dark world. Appropriately enough, the quarry from which we collected them was also profoundly gloomy, the muddy Ordovician rocks from which the trilobites emerged sooty black—and probably always had been from the time of their deposition. So we were finding black trilobites that lived in perpetual darkness in a dim quarry full of black rock. My eyes have never really recovered. Even odder, the other trilobites in the same rocks were huge-eyed swimmers. It did not take us long to deduce that the swimmers swam well above the lightless sea-floor on which the blind animals dwelt. They joined one another only in death. Eyes were lost because they were surplus to requirements. These trilobites had an affinity with blind cave crustaceans, new species of which are discovered every year: pallid creatures, in which pigment has bled from the body and light from the eyes. When they are brought to the surface, they look diseased, like tubers stored too long in the cellar. They are not ill, of course, merely drained of superfluities which are not necessary for their dim existence.

Blind trilobites were not confined to the deeper seas, although they were probably commonest there; others may well have been burrowers, for example. Nor would it be correct to think of them as degenerate. I was brought up in a tradition of thinking of animals in their geological history reaching some kind of acme, and afterwards—those that ultimately went extinct, in particular—having a phase of decline. There was, inevitably, the image of human frailty blended with this. Blind or otherwise specialized creatures might be described as degenerate, almost in the way of the Victorian black sheep of the family who contracted unnameable diseases and squandered the family fortune. I must confess that I once found something about this scenario attractive, and not merely because it dramatized episodes in the history of life in novelistic terms. It tied in with some inaccurate notions of Darwinian "fitness," too. Some animals were destined for (literally) blind alleys from which they never emerged. Other, fitter, ones prospered and brought forth evolutionary progeny. Curiously, it *is* true that I know of no example of eyes once lost and then regained. It is certain that trilobites primitively possessed eyes—their loss is always therefore a secondary adaptation; and, like virginity, sight can never be regained once it is lost. But the important point is that all blind trilobites were excellent citizens of their time. Sometimes they outnumbered their sighted cousins. In Shropshire, England, there are soft, greenish shales that crop out in the banks of Sheinton Brook, near the odd-looking hill known as The Wrekin, which are covered with the remains of the blind trilobite *Shumardia*, a perfect little miniature with only six thoracic segments and a glabella shaped exactly like the ace of spades. The same animals are abundant in Ordovician shales in Argentina; and I collected them again in southern China, equally in their hundreds. No dissipated roués these—they clearly spread around the Ordovician world in their millions. If there is a moral to draw from this it is that what we describe as adaptive success is also about the opportunities presented by the environment. Just as the ingenious inventiveness of the trilobite eye opened up a range of habitats so that they could hunt or swim in the open sea, even the loss of that same organ could qualify the animals to prosper in a sea-floor of soft mud. The glorious richness of the natural world is about multiplicity of adaptations, and it was so from the time of the trilobites.

The story of trilobite eyes I have related illustrates, appropriately enough, the principle of progressive illumination. This is the way science often works. As we learn more, we think of new questions to ask. We proceed from calcite to calculation to something like certainty. Imagination plays its part, of course, but our appreciation of vision is not, as Samuel Taylor Coleridge said, *A sight to dream of, not to tell,* for we try to convert our dreams into the solid stuff of published scientific articles. Advance in knowledge was dependent first upon recognition of the different kinds of eyes, and then of the ways they might have worked. At root, this is another form of taxonomy, another expression of humankind's talent for discrimination. Nor does the process come to an end. As I write this, a paper has just arrived from a Hungarian scientist suggesting that some trilobites may have had bifocals! I do not know yet whether this idea will pass the criticism of the next generation of investigators. What I do know is that there will be more to find out about trilobite eyes. That combination of imagination and rigour which is what science is all about has not yet probed out all the secrets: there is sight, and there is insight.

V

Exploding Trilobites

The play's the thing! As the theatre curtain rises we butt into the lives of the characters without preamble. There is no biographical account of the *dramatis personae* printed in the programme notes, describing their lives before they are presented to us on stage. We expect no more history than is given to us in the three hours or so that we occupy our seats. If it is a good play, as with any work of art, the experience will be satisfying within its own boundaries—we won't be left fretting to know about the early history of the characters, any more than we will demand that Macbeth overcome his remorse and next time triumph over his foes.

Drama is often used as a metaphor in the history of life. Animals have been described as "actors" on the ecological "stage." Mass extinction events that have interrupted more routine oscillations of fortune, the stuff of everyday evolution and decline, are "dramatic" interruptions in the "narrative." This kind of description engages the attention more readily than strictly scientifically-correct statements about "statistically significant elevations of extinction above background rates." Well, what's wrong with a little theatre? There will have been episodes in the history of life as dramatic in their way as that instant above the sea where Stephen Knight faced the trilobite and transformed his destiny. A long view of life does indeed see dramatic twists, where roles are reversed—as when mammals replace dinosaurs centre stage—or one old stager is replaced by another in a familiar part. I have heard it said that trilobites once played the ecological role now taken by crabs and lobsters, and most people with a passing knowledge of natural history will know that ichthyosaurs have been described as the porpoises of the Jurassic. We will let it pass that neither of these nostrums is exactly true.

When a play (particularly a whodunnit) is beginning to stall a little, one standard bit of theatrical "business" to bring it alive is to introduce an explosion. Boom! The audience jumps to attention; and, of course, under the cover of gun-shot it is possible to get away with murder, dramatically speaking.

Fossils of trilobites appeared suddenly in the geological record during the early part, but not quite at the base of the Cambrian period, perhaps 540 million years ago. If you are tempted by the word "dramatic" then this is one occasion where you could be forgiven for weakening. When you visit a rock section spanning the right bit of the early Cambrian—and there are such profiles in Newfoundland, Mongolia and Siberia—there will be not a sniff of a trilobite as you work your way upwards from one bed to its successor: this is the most methodical way to trudge upwards through geological time. Then, quite suddenly, a whole *Profallotaspis* or an *Olenellus* as big as a crab will pop out into your waiting hands as you split the rock. These are trilobites with lots of segments and big eyes: striking things, not little squitty objects. It is an appearance as dramatic as that of the sorcerer in *Swan Lake*, who accompanied the first theatrical explosion I ever experienced. You are tempted to cry out: "bang!" And as you continue to collect a foot or so higher into younger strata, the first trilobite will be joined by others, maybe half a dozen or so different species, and all individually distinctive ones at that.

More than a decade ago I examined the trilobitic early Cambrian rocks on the island of Newfoundland. Its Northern

Peninsula sticks up as stiffly as an abusive finger on the western side of the island. There is only one road running northwards up the Peninsula* from Deer Lake to St. Antony. After less than an hour on this road you enter the Gros Morne National Park. The road winds alongside Bonne Bay, a beautiful, huge, drowned inlet into which the surrounding hills plunge somewhat recklessly. The valley which is now occupied by Bonne Bay was long ago excavated by a river, but the rise in sea level at the end of the last Ice Age drowned it, so that now the sea probes impertinently into its every corner. Firs and aspen and birch vie with scrubby alder to make the inland almost impassable on foot. Blackflies and mosquitoes are further discouragements. The road loops and curves, not allowing any driver to take his eyes off it for long. Strata have been sliced through to make the road passable: they were folded in such a way that they now plunge at various angles into the sea. On a great, flat brown bedding plane of rock some foolish exhibitionists have painted their initials a foot or so high in white paint: "RW luvs SDM." What they should have written is: "Amazing trilobites found here!" But any such search would have to be frustrated, because Parks Canada have strict rules about collecting by permit. If you do have permission, as I did, you can break open shales to your heart's content, and if you are lucky you will be rewarded with a big, juicy *Olenellus,* or you may find fragments of another trilobite like *Bonnia*—presumably named after the Bay. On the other hand, you may well discover completely different kinds of fossils: I recall that I found a fine specimen of a very primitive echinoderm (the group which includes starfish and sea urchins) called *Protocystites walcotti*—no prize

---

*Newfoundlanders are often humorous and hospitable people. They have been used as the butt of jokes by mainland Canadians, allegedly for their lack of sophistication. The only instance I can recall of a real "Newfie joke" was when I was leaving the Gros Morne Park in search of food. At the inn at Cow Head a laboured hand had scrawled a notice in jagged chalk letters: "Cow Head MOTEL best food on the whole PENISULA."

for guessing after whom it was named. Nearby, a small mollusc turned up.

The early Cambrian level is like this almost everywhere. A variety of shells can be found, some of which can be placed with their living relatives in one or another of the great "phyla"—the major divisions of the animal kingdom, like Mollusca, Arthropoda or Echinodermata. The Cambrian strata which lie below those including the first trilobites often yield a variety of small shells, tubes, plates and nets sculpted into a host of shapes, many of which defy ready recognition as any part of an animal we know about. They are collectively referred to as "small shelly fossils." It has recently been shown that the variety of these may be more apparent than real, because Simon Conway Morris and John Peel have proved that many different "small shellies" belong together to make up a much larger, plated organism, called *Halkieria*. The little plates may be no more than the components of a kind of chain mail. But what is real enough is the sudden appearance of shells—hard parts, skeletons, call them what you will. Animals contrived to use minerals to support their soft anatomy over (geologically speaking) a very short period of time at the base of the Cambrian. Not only that, there is now a number of fossil faunas from the Cambrian that exceptionally preserve animals *without* hard skeletons—those soft-bodied animals which are generally so rare in the fossil record.

The most famous of these localities is the Middle Cambrian Burgess Shale in British Columbia, arguably the most illustrious fossil deposit anywhere in the world, thanks to S. J. Gould's account of its treasures in *Wonderful Life* (1989). But fossil faunas have now been recovered from even earlier—Lower Cambrian—rocks which are almost as diverse. Those discovered in Chengjiang in China and Sirius Passet in Greenland are the most spectacular. They all confirm the same idea: that a great diversity of life was already present in the Cambrian period. Some of these animals had hard exoskeletons or shells, others did not. Some were familiar, others strange and

puzzling. There were assuredly more arthropods than anything else—nobody could mistake those jointed legs. But
what an extraordinary bounty of articulated phantasmagoria
were propped upon those typically spindly legs! I cannot here
elaborate on all the other odd animals in these faunas: I will
only reiterate that, like the devils that persecuted Christ's
madman, they were legion. Gould even famously, and erro-
neously, argued that there was more "diversity" (or, not to
misrepresent him, what he termed "disparity") in the Cambrian than was seen ever again in the subsequent history of
life.

The curtains had been drawn on a vital drama, and up on
stage there was an elaborate cast, all dressed up to the nines in
both familiar and unfamiliar garb. It was a *coup de théâtre* that
could not have been bettered by Peter Brook. Dazzled by the
brilliant array of costumes the onlooker is confused and over-
awed by the sudden burst of spangles and tulle. Surely this is
the ultimate show, a more heterogeneous parade than could
have been dreamed by a zoomorphic imagination in the grip
of hallucinogens (why, one animal is even called *Hallucigenia!*). This was a show without precedent, for, like all shows,
the cast had no history, their stage presence was a portrayal of
the moment; a sudden, glorious, opening night. There was an
explosion of characters after a dearth of drama.

Where the theatrical analogy fails is that, unlike a stage
play, the history of life *does* demand a before and after, a
beginning and (in some cases) an end. We can be stage struck
for only so long, before we try to peel off the make-up, and
strip away our initial astonishment to determine the substance of the characters. Life has a biography that is rooted
back in time, because all animals are ultimately descended
from a common ancestral species, an evolutionary Adam. This
accounts for the shared genetic legacy of all creatures, be they
ever so great or ever so small.

The sudden appearance of fossils of so many kinds at this
geological moment has become well-known as the Cambrian

evolutionary "explosion." The dramatic metaphor is no coincidence. It embodies the idea of a chain reaction somewhat out of control: that is, a hugely elevated rate of evolution. And, as in all explosions, the bomb is small but the effects are disproportionate. But this is a creative explosion, rather than a destructive one, so that when the curtains of the fossil record are drawn back suddenly the whole bang-shoot is revealed in the spotlights after the chaos of creative "explosiveness." And why the necessity for explosiveness at all? This is because virtually the only fossils found below the Cambrian in the late Precambrian, or Vendian, strata are either simple plants and bacteria, or else soft-bodied oddities—known as the Ediacara fauna—which are difficult to interpret as any kind of ancestors of what was to follow. The earlier stage was strangely lacking any characters that we could recognize. It is as if the players in the Cambrian drama had appeared from somewhere else, having dressed and made up in secret. Where are the late Precambrian brachiopods, molluscs, echinoderms and arthropods that might have provided the Prologue?

I must confess to a certain diffidence in writing about the Cambrian "explosion" question. It has excited so much passion and disgreement among its several interpreters, some of it remarkably ill-tempered, that I doubt whether it is wise to walk into this volatile environment without a safety helmet. I recall the words of Charles Darwin in his *Autobiography:* "I rejoice that I have avoided controversies, and this I owe to Charles Lyell,* who . . . strongly advised me never to get entangled in a controversy, as it rarely did any good and caused a miserable loss of time and temper." However, the trilobites demand it. It might seem strange to the reader that events which happened more than 500 million years ago are still capable of causing the secretion of abundant venom in living organisms. It is undeniable that some of the most obvi-

---

*Sir Charles Lyell, whose *Principles of Geology* was a profound influence on the young Darwin.

ous "explosions" are to be heard from the proponents of one theory or another. The puzzle of the sudden appearance of shells at the base of the Cambrian has been with us for a long time. Darwin certainly recognized it; in Chapter 9 of *On the Origin of Species* (1859) he had despaired that "the case at present must remain inexplicable . . ." One hundred and forty years later there is no shortage of explanations: there is just a shortage of agreement. I am compelled to join the fray because, of course, trilobites were among the observers of the explosion—if indeed there was one—and were hardy survivors from the Cambrian into the Ordovician and beyond, outlasting many of those animals, termed Cambrian evolutionary "experiments," which perished without benefit of descendants. Because trilobites appeared alongside the first arthropods in the Cambrian they must be part and parcel of the "explosion"—caught in the crossfire at the very least.

One of the Burgess Shale fossils is a trilobite called *Olenoides*. It is another of those rare species in which the limbs can be clearly observed: the peculiar preservation in the Burgess Shale allows us to see shiny impressions of legs and gills, and even guts. Like *Triarthrus, Olenoides* (fig. 7) has been examined by most of the leading trilobite specialists from the time of its discoverer, Charles Doolittle Walcott. It will come as no surprise by now that the definitive description of the animal was made by Harry Whittington in the 1980s. In its essentials, the trilobite had a similar arrangement of limbs to that I have already described for other species. On the cephalon there were three pairs of limbs and flexible antennae; each thoracic segment also had the usual biramous limbs. One difference from the usual pattern was that *Olenoides* also had a pair of special antenna-like appendages at its rear end, known as caudal furca. But the basic trilobite design was confirmed yet again. Harry noticed that the bases of the walking legs were truly massive and equipped with sharp spines which faced inwards towards the centre line of the animal. He interpreted this spiny corridor as an area where prey items were

Olenellus, one of the earliest trilobites from the Lower Cambrian strata: already a specialised creature.

shredded and passed forwards to the mouth at the back of the hypostome: *Olenoides* was a predator well able to gobble up a variety of "worms"—which also abound in the Burgess Shale. Predator and prey made their debut together.

Trilobites were one of the many animals which were so dramatically exposed when the curtain was lifted on the Cambrian Burgess blockbuster. But for many years before the discovery of the Burgess Shale by Walcott in 1909, trilobites were almost the *only* arthropods known from the Cambrian strata, preferentially preserved as they were by virtue of their calcite exoskeleton. They had come to be a surrogate for all that was primitive among joint-legged animals. It was accepted on the nod that trilobites could stand in for an arthropod ancestor. Even early observers could clearly understand that they were quite complex animals, eyes and all. How then to account for their sudden appearance? Charles Darwin was unusually confident in the *Origin of Species:* "I cannot doubt that all the {Cambrian} trilobites have descended from some one crustacean which must have lived long before the {Cambrian},"* he wrote, thirteen years before Thomas Hardy confronted his hero with another such "primitive crustacean." The attribution of trilobites to the arthropods may be almost instinctive. The anthropologist Kenneth Oakley made known a perforated specimen, a pendant probably, recovered from the Grotte du Trilobite in Yonne (France). This is a late Palaeolithic cave, and records the first interaction between mankind and trilobite. In the same cave there was found a beautiful carving of a beetle. "It does seem reasonable," says Oakley in 1965, "to infer that the trilobite would have appeared to the untutored yet observant and thoughtful Magdalenian as a kind of insect in stone." Quite so. The Magdalenian saw an insect, Darwin a crustacean, Walcott (eventually) an arachnid, that is, a relative of the spiders and scorpions—they cannot all be right.

But an objective answer to the simple question: what is the trilobite's closest relative? has proved rather elusive, and is, like it or not, intimately bound up with the question of the Cambrian evolutionary "explosion." In the first place, the wealth of different arthropods now known to be present in the Cambrian robbed the trilobites of their right to claim exclusive primitiveness. Any one of these other animals might have a prior claim. Several of them also had the kind of branched limbs with which we are now familiar. "Tracks" scraped by such limbs are familiar as fossils, even predating the appearance of the body fossils themselves by a short time. It had been assumed that these tracks—which carry the names *Rusophycus* and *Cruziana*—were made by trilobites, but that was

---

*The original text reads "Silurian" where I have written {Cambrian}. At the time Darwin wrote the distinction between Cambrian and Silurian strata had not yet been recognized (see Chapter 2).

when trilobites were the only known possible culprits. Now there were many other possibilities. So the discovery of the Burgess Shale and other, even earlier Cambrian fossil soft-bodied faunas made everything more complicated.

Notwithstanding, I shall try a simple explanation.

When Harry Whittington and his research students Derek Briggs and Simon Conway Morris (now famous professors in their own right) minutely studied the Burgess Shale animals in the 1970s and 80s they tended to emphasize the peculiarities of the fossils they studied. They were, after all, describing the costumes of spectacular performers for the first time in detail since Walcott had pulled back the stage curtains.* The distinctiveness of the limbs of some species, carapaces of others, uninterpretable features of yet others seemed to support a notion then current that arthropods (and by extension other kinds of animals) had appeared from more than one ancestor. This is correctly described as "polyphyletic" origin.

At the height of the popularity of "explosivism" polyphyly was rife. Harry Whittington once believed that the various kinds of Burgess arthropods arose separately from different soft-bodied ancestors in the Precambrian. At its extreme, this view had it that some of the Cambrian animals were so different in design that they merited being placed in the highest category of animal classification—a phylum. They were not molluscs or arthropods, or whatever, at all, but merited a separate phylum of their own, or so it was alleged. Simon Conway Morris is famously quoted as opening a drawer, encountering a new fossil animal, and exclaiming, "Oh fuck, not another new phylum!" He has probably had cause to rue these words. At a more modest level, awkward arthropods that showed peculiar characteristics were regarded as repre-

---

*I should mention that a number of previous professors have had their turn at the Burgess animals. The arthropod studies of Percy E. Raymond of Harvard University in the 1920s and Lief Størmer of Oslo in the 1940s contributed much of interest to the story of their interpretation, notably the recognition of the "trilobite-like" limbs of these animals.

sentatives of "failed designs." There was no shortage of these curiosities: arthropods with giant frontal appendages, or great feathery antennae, or huge numbers of segments. All were supposed to have appeared as a result of a special burst of evolutionary creativity at, or just before the base of the Cambrian, 545 million years ago. This was the "explosion." The dramatic appearance of a plethora of life on the geological stage was considered a true measure of evolutionary history. Trilobites were just one of many designs thrown up at the same time, but they were undoubtedly part of it, and must have observed their idiosyncratic spindly or bristly neighbours through their unique crystal eyes. No other Cambrian "experiment" mastered that particular optical trick.

When S. J. Gould explained an early version of the explosion theory in *Wonderful Life,* he described the various animals and laid out the conclusions he drew from them. He attributed, with some generosity, much of the novelty of the interpretation of Cambrian events to Simon Conway Morris; "as for so much of this book, I owe this example to the suggestion and previous probing of Simon Conway Morris" (p. 293) was a typical endorsement. The redescription of the Burgess Shale fossils was a team effort overseen by Harry Whittington. Different beasts were studied by Conway Morris, Derek Briggs, David Bruton and Chris Hughes. I had recently gained my first employment as a trilobite specialist when the "Burgess boys" were ensconced in their offices in the Sedgwick Museum, Cambridge, where they spent all day, every day, feverishly preparing and photographing and discussing their marvellous animals. I was a fascinated bystander who participated in the conversations and speculations as they happened. I pored with Derek Briggs over fossils of the arthropods *Sanctacaris* or *Canadaspis* on their quotidian wooden trays, containing slabs of black shale which looked so ordinary yet carried on their surfaces such extraordinary objects. From the outset, I was interested in how the newly interpreted animals would cast light on the affinities of trilobites. Curiously, I do not remember hearing the word "explosion" once in those early days.

As with most new and attractive theories, it was not long before all manner of other sources of evidence were recruited in support or explanation of rapid evolutionary turnover at a critical time. We met Hox genes in the last chapter: those genes which control aspects of the sequence of development in all animals. Arthropods are typically composed of "packages" of segments—the reader will by now be familiar with the trilobite's own arrangement of cephalon, thorax and pygidium. But these packages are differently arranged in the various major kinds of arthropods: the head may contain different numbers of segments, as may the thorax. They are like trains buckled together with different arrangements of carriages and cars. One theory suggested that the Cambrian "explosion" might record a critical period in the expression of Hox genes, which at that time, uniquely, were "switching on" new arrangements of batteries of limbs and segments. The Hox genes functioned like the presiding genius of a biological marshalling yard, mastering new rolling stock. In the grand democracy of the early Cambrian a swarm of these creations could survive, and devil take the hindmost. Some would produce evolutionary progeny, others would fail after a geological moment of improbable fecundity. Yet another theory invoked a phase of doubling up of lengths of the genetic code at this creative time, a genetic change which increased the possibility of innovation and variation in body designs. It seemed for a while that Darwin's "inexplicable case" might be solvable after all. The "explosion" was a special moment in time—triggered, possibly, by some environmental threshold that had been crossed in the Cambrian world—when the possibilities of life gloriously expanded, and when parts in the evolutionary play became suddenly multifarious. For a while, any strange role was permitted. In popular accounts it became an ancient moment of madness, a magnificent evolutionary Mardi Gras, when a parade as bizarre as could have been

devised by a surrealist on speed would be permitted for a geological day. "See the crystal-eyed monster!" "Roll up, roll up,
for the shiny, tubiferous wiggly orphan *thing* with no relatives!" The freak show was open for trade.

It seems almost a pity to spoil the show, to strip away the
dressing and reveal the actors beneath. Most of us prefer
glamour to analysis—glamour can be enjoyed with a smile
and a cheery wave, whereas analysis requires thought and
intellectual work. But it is necessary to do this critical work, if
only to understand the place of the trilobites in the Cambrian
charivari.

From the first there were a number of scientists who
doubted the account that Steve Gould had presented, however much they admired the manner of its delivery. I was one
of the doubters myself, ever since I reviewed *Wonderful Life* in
the magazine *Nature* shortly after its publication. At the time I
had already started an attempt to look at the Burgess animals
and their relatives in a different way.

This work was done jointly with Derek Briggs, who was so
familiar with the details of the Burgess arthropods. Instead of
emphasizing their peculiarities, we were looking for the significant similarities they *shared* with one another. The technique of analysis we used is called cladistics. Although its
details are technical, the main principle of cladistics is very
simple: it is an attempt to classify organisms on the basis of
evolutionarily derived characteristics. To give an easy example: if a cladistic analysis were made of a shrew, an elephant
and a lizard, the shrew and elephant would share several
characters, such as uterus, mammary glands, warm-blooded-
ness and (patchily in the elephant, admittedly) hair, which
would readily identify them as being more closely related to
one another than to the lizard. Both shrews and elephants are
mammals: we don't imagine that complicated characters like
mammary glands arose on more than one evolutionary occasion. On the other hand, that both shrews and lizards eat
insects while elephants digest trees is not an indication of

biological affinity at all, but an expression of adaptations for earning a living. Nor does the peculiarity of the elephant, its ineffable trunk, make it a mammal, or help us to decide whether or not it was more closely related to the shrew than to the lizard. Cladistics identifies only the significant advanced characteristics as the basis of classification, while similarities which are just a reflection of an earlier, shared history do not figure. The four limbs of lizard, shrew and elephant reflect the fact that all belong within a greater group—the tetrapods— with an ancestry stretching back to the Devonian, and do not contribute to solving the particular problem in hand.

So Derek and I had the job of enumerating all the characteristics of the Burgess Shale arthropods which were present on several animals—features of the legs, or the number of limbs incorporated into the head, for example. From the distribution of these characters we wanted to draw a tree diagram of the relationships of the fossils. As in any family tree of the Cholmondely-Smythes or the House of Windsor, we should be able to see who is most closely related to whom and how the more distant members of the family slot in. Except that we were dealing with common ancestry, rather than pinpointing Uncle Ernest as an actual ancestor. The more animals you study, the disproportionately more *possible* ways of arranging them into trees there are; so you soon need a computer programme to sort out the best arrangement, especially if some characters may genuinely have appeared more than once during evolution. "Best" is, of course, a charged word— how do you know what is best?—and cladistics programmes have various ways of deciding this, but most boil down to a preference for the simplest tree. In the late 80s most scientists interested in this technique were using a programme called PAUP (a simple acronym for the intimidating Phylogenetic Analysis Using Parsimony) developed by an American from Illinois, David Swofford. In evolutionary circles Swofford's name is about as well known as Hawking's in astrophysics.

We were stripping these Cambrian arthropods—including

the trilobite *Olenoides*—down to their essentials to see if they would "fit" into a tree of descent. If some of the "oddball" animals could be accommodated convincingly into the tree, then it would mean that their distinctiveness might have been over-emphasized by the explosionists. They might have been too easily dazzled by their spectacular motley, to the extent that they failed to see that underneath they shared a common dress.

What surprised us was how *easily* we could produce a tree relating nearly all of the Burgess Shale arthropods. This was the first time that such an objective tree of descent had been produced, and one of the most interesting things to us was that the trilobites were quite high up within it. So much for their traditional role as the archetypal primitive arthropod! If they had been primitive then they would have been somewhere near the base. Suddenly the distinctiveness of the crystal eyes seemed to make more sense. Then the idea that the different arthropods arose independently was knocked severely by the fact that a perfectly reasonable tree of descent could be reconstructed, with all the arthropods having descended, ultimately, from a common ancestor. Some of the strange arthropods from the Burgess Shale were truly no stranger than trilobites—it is just that we have had a hundred years or more to get used to trilobites. Familiarity breeds, well, familiarity. If there was an "explosion" then it was a remarkably orderly one. In detail there were many problems with our aboriginal tree, which was very much a first attempt—and crude, as such attempts always are. But in the ten years that followed other people have tried their own versions, including our own friend Matthew Wills, and many of the original features have been preserved. In other words, our tree must have had a germ of truth.

Derek and I thought we would publish this tree as a paper in the journal *Nature*, but *Nature* had other ideas. The reader should know that publication in a scientific journal is no simple process. You must submit the manuscript to the journal, obeying to the letter all caveats about format and length. Then the journal will send the paper out to referees, and if it is *Nature* you will probably have the pickiest referees in the business. In the majority of cases they will recommend "rejection." Only geniuses like Richard Feynman or Stephen Hawking routinely get a "yes"—everyone else endures some degree of pain. No first-time novelist suffers more when the terse note comes back, saying "The editors regret . . ." So you can imagine our level of disgruntlement when the paper describing the Burgess tree was given the bum's rush by *Nature*. We licked our wounds and resubmitted our paper to the North American equivalent of *Nature*—*Science* (New York)—which is about the only scientific journal with an equal reputation. To our relief, after a month or two's agony, the little paper was accepted and was published in 1989.

Since then a great deal more has been learned about the fossil faunas that preceded the Burgess Shale in even earlier Cambrian time. It has been made abundantly clear that many of the arthropods that were described from the Burgess Shale now have relatives in older Cambrian strata. From China, the Chengjiang fauna has yielded many beautiful animals. The story of their description makes the Burgess Shale controversies seem almost decorous. Rival teams of collectors have been racing to be first. Peasants have been paid to come up with the goodies, even whisking fossils out from under the noses of their competitors. There have been rival publications. Cloak and dagger has been followed by backstabbing. Chen Jun-yuan and his team of westerners have tried (with some success) to race ahead of Dr. Hou and *his* alternative team of westerners. Sometimes, you don't know which name you should use for a given fossil, Chen's or Hou's. Greg Edgecombe, an incorrigibly amiable naturalized Australian, who has done much to make these animals known to the world, whistles through his teeth when I mention future Chengjiang visits to him. "Never again!" he says. "Not f****ing likely!" There is something about these ancient fossils that excites four-letter words.

Science, of course, does not give a fig for these examples of internecine warfare. The truth will out, and it matters not if it is at the cost of bruised egos or skulduggery. Some time in the next decade or two all the vested interests that have fought over Chinese fossils will seem as tragi-comic as the nineteenth-century battles between Professors Marsh and Cope to name the greatest numbers of dinosaurs in America. As far as the "explosion" is concerned, the continuation of evolutionary lines back into strata earlier than the Burgess Shale simply increases what Gould later called the "intrigue and mystery" of events at the base of the Cambrian. If you add to the brew the Briggs/Fortey tree (or one of its better subsequent versions) there is a very simple question to be asked. If the different kinds of arthropods extend to the early Cambrian (including trilobites, which are near the *top* of the tree), then must it not be the case that the only time for the still earlier branches on the tree to have split is within the Precambrian? And since these arthropod branches will connect in their turn with still more and deeper branches in the branching history of the whole of animal life, this takes us somewhere still older and more profound again. You cannot have a great-grandson without a great-grandfather. We have already met this argument with regard to the history of eyes in the last chapter. Eyes are deeply implicated in the history of life. The eyes of trilobites are tied by a genetic bond to other eyes in other animals, all the way back to the first simple eye-spots. The estimates of divergence between major animal groups based on molecular clocks (which have faults, admittedly) is somewhere between about 1000 Ma and 650 Ma, but both are substantially before the Cambrian at 545 Ma. Maybe that dazzling opening scene of animal life blinded us to a modest, and much longer earlier drama after all.

Some years ago I made a straightforward observation

about the very earliest trilobites. When they first appear in the Cambrian strata, they are already different in various parts of the world: not just different species but different genera— even families. If you go to China you will find specimens of a compact little trilobite called *Parabadiella*—but not *Olenellus*. If you go to New York State you will find *Olenellus* and its friends—but no close relatives of *Parabadiella*. If you go to Siberia along the Lena River in the mosquito-ridden summer you will see the best exposed Cambrian in the world, but the first trilobite you find will be yet another one, *Profallotaspis*. Since all trilobites probably descended from a single ancestor it seems obvious that we must be missing a phase of their fossil record—the time required to evolve their different forms in different areas. It all points to some considerable history missing from these particular rock sections at the base of the Cambrian. This is proved beyond doubt in those beautiful Lena sections where you can see the effects of erosion before the appearance of the Cambrian fossils. Is this erosional phase the time for the trilobite to be "descended from some one crustacean . . . before the beginning of the Cambrian" described by Charles Darwin?

One thing of which we may be sure is that the trilobites were *not* descended from any sort of crustacean—trilobites and the arachnid horseshoe crab *Limulus* (p. 158) shared a common ancestor, which in its turn shared a common ancestor with the crustaceans. Trilobites were distant cousins, not progeny of the crustaceans. But it is also true that many of the fossils that fit in low on the tree of descent of the arthropods shared features that once upon a time would have been thought of as typically trilobite. Those limbs with two branches that C. D. Walcott laboured so hard to reveal turn out to be very common among all manner of Cambrian soft-bodied arthropods, too. Arthropods leading to crustaceans probably had them—as did those that would lead on to *Limulus* and scorpions. In a word, biramous limbs are primitive. Any one of these animals could have scraped simple trails like the ones found in the earliest Cambrian of all in eastern Newfoundland. Some other facts are also becoming clearer. The closest relatives of the typical arthropods are some little animals with stumpy legs known as velvet worms (Onychophora). They still live today—mostly under rotting logs in warm, damp climates. In the Cambrian they were much more numerous and varied, but also submarine. Graham Budd has shown how a number of really weird-looking creatures are velvet worms at heart. So is the once nonpareil oddball from the Burgess Shale, *Hallucigenia*. It is a measure of how misleading Gould's theory would have been if taken at face value: these animals which were once touted as designs of uncommon originality would have been simply labelled "failed experiments" and there's an end to it. As it now is, they have been recognized as important steps in our understanding of the subsequent history of life. The lesson of cladistics is that it is what animals share that is important in identifying their relatives, not our subjective judgements about oddity. We must focus on the elephantine womb, not the trunk, if we want to place the pachyderm in the scenario of Nature.

So now we are left with a paradox. There is a tree of descent which helps us understand the history of our characters before their spectacular appearance on stage—but of this earlier history there is no evidence. Even the traces left by animals, their scratches and burrows, are rare before the latest Precambrian.* Where can the animals be? Either all the rates of origination must be speeded up beyond comprehension in the "explosion"—and out of this "explosion" a variety of trilo-

---

*As this is written there are new reports from India of much older tracks and trails, in rocks up to a billion years old or more. There is little doubt that these markings were made by animals, and indeed it is a scandal that previous reports by Indian geologists and published in Indian journals have hitherto been ignored. There is however some reason to question the accuracy of the dating, and for the moment final judgement must be suspended.

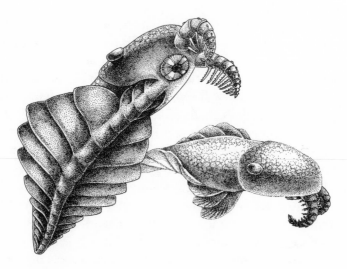

Anomalocaris, at first claimed as a "weird wonder," but now known to be related to primitive arthropods and hence to trilobites.

bites must crawl alongside everything else—or there must be some other explanation. Like T. S. Eliot's Mystery Cat:

But when you reach the scene of crime McCavity's not there!

The explanation of missing time might apply to some rock sections, like the Siberian ones, but it does not work for eastern Newfoundland where the rocks record a complete narrative. My favoured theory is that the earlier branches in the tree were tiny animals, which were not easily preserved as fossils. It is not necessary to be large to be a perfectly good arthropod (or mollusc, come to that). The sea swarms with tiny arthropods today that have left no fossil record. I like to quote the tiny copepods, which are members of the plankton so numerous that they can turn the seas black. Yet almost their only fossil is a species preserved in the body of a fossil fish. Were it not for their miraculous preservation in amber our knowledge of past insects would be dreadfully inadequate (as it is, we know from amber hundreds of the most delicate of all, mycetophyllids, the fungus gnats, so delicate in life that a puff of wind destroys them). What happened at the base of the Cambrian was probably as much an increase in size as a sudden appearance of new types of animals. This may well have been a genuinely rapid change. We know from many fossils that increase in size is quite an easy goal in evolutionary terms. Mammals, for example, seem to have undergone a very rapid increase in size after the demise of the dinosaurs 65 million years ago. It even seems possible that the same size increase allowed the possibility for the secretion of shells. Muscle support becomes much more crucial when an animal reaches a certain critical size. So the "explosion" was a dramatic appearance of characters that had been rehearsing out of sight for more than a hundred million years.

The explanation just given holds out the possibility of discovery. Maybe one of the readers of this page will discover the Precambrian equivalent of amber. Very recently, a late Precambrian animal embryo was discovered in China, amazingly preserved cell by cell in the mineral calcium phosphate; age by itself is evidently no proof against miracles. It would be wonderful to amaze the world with proof of the missing stages of evolution, tiny animals that set the designs for the future of life. Somewhere, there should be a small trilobite, an animal with the potential for spinning an almost endless variety of costumes for three hundred million years. The search continues.

This is not quite the end of explosions.

There have been several more accounts of the Burgess Shale and the Cambrian since *Wonderful Life*. Most of the arguments about the truth or otherwise of the "explosion" of phyla have been in the pages of scientific journals, in which a convention of decorousness is observed. *Toujours la politesse!* Steve Gould knew that I did not agree with his conclusions, but this made no difference to the cordiality with which we could meet: we could wave across a conference room without any gritting of teeth. I do not suppose he was tempted to take a small model of me and stick pins in it, any more than I was tempted to steal some of his personal effects and cast a hex. Scientists seldom do things like that: what they are chiefly interested in is the advancement of truth. Richard Dawkins tells a good story about a senior professor coming on stage to shake the hand of the young scientist who has just disproved the old man's most cherished theory: the older man earns a standing ovation. Well, that is how it is supposed to be, according to the etiquette books.

The Smithsonian in Washington, DC, mounted a Burgess Shale exhibition, where the public can see the extraordinary beasts for themselves: the accompanying literature is unobjectionable, and factual enough. At about the same time as this exhibition was being opened, the extreme of "explosiveness" emerged in a book by the two McMenamins, Mark and Dianna—professors in a small East Coast university—called *The Emergence of Animals* (1990). In this work they claimed up to a hundred* animal phyla "exploding" into life in the Cambrian; most of them are also claimed to have died out, leaving no progeny. They popped up like so many jack-in-the-boxes, and then auto-destructed, in the manner of some post-Dada extravagance designed to outrage. This view out-Goulded Gould ten-fold. The extraordinary thing to an objective reader is that there is no attempt to justify why these hundred or so "Cambrian phyla" should be recognized as one of the great divisions of the animal kingdom—how, exactly, do they differ from each other so much that they "deserve" to be called a phylum? Not a word. How do so many of these separately evolved creatures share curious similarities unless they were descended from a common ancestor? And if they were so

---

*Most modern textbooks list about thirty phyla in the living fauna, which embraces all the extraordinary diversity of organisms. Each phylum represents a fundamentally different design in anatomical organization. Thus the McMenamins identify at least a three-fold richer world in the Cambrian.

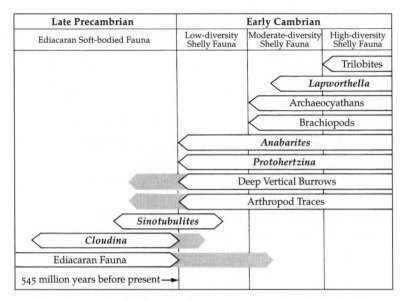

The place of trilobites in the emergence of animals in the early Cambrian, after McMenamin and McMenamin, *The Emergence of Animals* (1990). But where are the ancestors of the trilobites?

descended—then surely they belong to the *same* phylum? Not a scintilla of discussion. One is driven to the conclusion that these particular writers regard it as only necessary to appear on the Cambrian stage dressed in any sort of odd costume to be called a phylum. The time of the performance alone is a guarantee of novelty.

Nearly ten years after *Wonderful Life* appeared another book made an even more explosive sally into this arena. This time it was written by the star of the original Cambridge *enfants terribles*—and the hero of Gould's Cambrian *Weltanschauung*—Simon Conway Morris. In the ten years or so since Steve Gould transcribed the significance of the Burgess Shale for the world (at least, his view of what was then understood in Cambridge), Simon had had plenty of opportunity for second thoughts. His revised view now is apparently like that I sketched earlier: a rather defused "explosion." Simon both accepted the need for an earlier history of animals and rightly pointed out the ways in which the Cambrian remained a distinctive period, when shells appeared, genuinely rapidly, alongside good fossil faunas of animals that lacked them. There was nothing very incendiary here. I would say that Simon had come around to seeing the Cambrian faunas in their context at a crucial phase in the genealogy of life. The explosions were reserved instead for Stephen J. Gould. I have never encountered such spleen in a book by a professional; I was taken aback. Gould doesn't write, says the author, he produces "perorations." He lacks originality, while laying claim to it. This little passage from *The Crucible of Creation* (1998) will give something of the flavour: "Again and again Gould has been seen to charge into battle . . . strangely immune to seemingly lethal lunges . . . Gould announces to awestruck onlookers that our present understanding of evolutionary processes is dangerously deficient . . . We look beyond the exponent of doom and there standing in the sunlight is the edifice of evolutionary theory, little changed." This is a rather gassy way of saying that Gould is a mountebank.

It is one of humankind's less attractive foibles that success breeds envy, and since there is probably no one in biological science to rival Steve Gould in worldly and critical success—at least among the literati—it is not surprising that some of his rivals for the spotlight focus their attention upon him. It is, of course, perfectly legitimate to have differences of scientific opinion—in fact, it is an essential ingredient of progress. But what surprised me here was the unwonted explosiveness, the bilious ballistics. The detail of the attempt to cast Gould in a poor light extended into the depths of footnotes. Gould (and R. C. Lewontin) wrote a famous paper in 1979 with the rather overblown title "The spandrels of San Marco and the Panglossian paradigm: a critique of the adaptationist programme." But it addressed an important point about whether all structures found in Nature *had* to have a purpose. In one of his virulent footnotes Simon Conway Morris takes Gould to task for architectural inaccuracy—apparently the structures in San Marco should not be called "spandrels" at all! Tsk, tsk—as if such a terminological pinprick could puncture all the inflations of the paper. Such hypercritical zeal has to well up from a deep source. Why should Simon wish to bite the hand that once fed him? If you look at those little silvery fossils in their neat trays it is hard to believe that they can be the origin of such dispute; nor should trilobites and their allies take responsibility for any verbal bombardments. Conway Morris and Gould subsequently slugged it out in the pages of the magazine *Natural History*. I do not subscribe to the cynic's view that such disputes are part of the "hype" to increase book sales—such antipathy cannot be faked. I was reminded of a ballad by Bret Harte ("The Society upon the Stanislaus"), describing a nineteenth-century fracas in a scientific society over—what else?—fossil bones:

Now, I hold it is not decent for a scientific gent
To say another is an ass—at least, to all intent;
Nor should the individual who happens to be meant
Reply by heaving rocks at him to any great extent . . .

In less time than I write it, every member did engage
In a warfare with the remnants of a palaeozoic age;
And the way they heaved those fossils in their anger was a sin,
Till the skull of an old mammoth caved the head of Thompson in.

I could only diagnose the cause of Simon's ire as being the very praise that Gould once heaped upon him. To return to Richard Dawkins's story, this is like the young professor stamping hard on the foot of the older professor. *Wonderful Life* was such a global success. There, preserved in the aspic of a print that could never be unprinted, was the Conway Morris of "oh fuck! not another new phylum!"—the Conway Morris of the early 1980s. The nineties version disowned the ideas of

19. *Crotalocephalus*. The glabella has become completely crossed by strong furrows. The thoracic tips are produced into long spines, as is the pygidial margin. The lateral view shows how the thorax can arch up, and shows the convex eye lobe. Devonian, Morocco.

20. **RIGHT**
Acanthopyge, a trilobite related to *Lichas*. The size of a crab, with a very peculiar glabella and a pygidium which is larger than the head-shield. Devonian, Morocco.

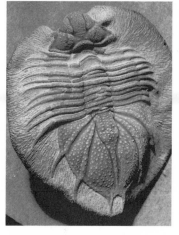

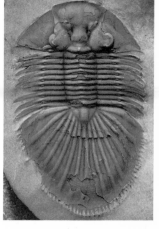

21. **FAR RIGHT**
A relative of *Scutellum*, *Thysanopeltis*, with an enormous, fan-like pygidium which is much longer than the head shield. Specimens are often 10 cm long. Devonian, Morocco.

22. A cluster of five *Cyphaspis* from the Devonian limestones of Morocco; very similar trilobites can be found worldwide. This particular species had a pair of "devil's horns" on the glabella. A long spine on the thorax may have helped the trilobite right itself if it landed on its back; the pygidium is comparatively small.

23. One of the survivors among the trilobites, *Griffithides,* from the Carboniferous (Mississippian) rocks of Indiana. Specimen about 5 cm long.

24. *Paraharpes,* a remarkable animal in which the genal spines are prolonged into a "brim" which extends alongside the length of the trilobite. The outer part of the brim is flattened for resting on the sediments. This trilobite has very degenerate eyes and numerous thoracic segments (length usually 5–6 cm). Ordovician, Scotland.

25. A close relative of *Ampyx, Cnemidopyge,* from the Ordovician shales of central Wales. A blind species, which carries a spine on the middle of its head like a rapier. The genal spines are equally long (the one on the right is shown here) and extend backwards beyond the body of the animal, which has six thoracic segments and a strongly furrowed pygidium.

26. Three blind trilobites. A group of three specimens of *Conocoryphe* from the Cambrian of Bohemia (modern Czech Republic) made famous by Joachim Barrande. Two of the specimens are preserved right way up, the third on its back. Note the comparatively small tail, and the fourteen thoracic segments. This is one of many Cambrian trilobites with a "flowerpot-shaped" glabella.

27. ABOVE One of the "strawberry-headed" trilobites of the Silurian period, *Balizoma variolaris,*
beautifully etched out in natural relief. The
pygidium begins after the twelfth thoracic segment. The head is richly covered with coarse
tubercles. From the Wenlock Limestone
(Silurian) of Dudley, Worcestershire, UK.

29. *Selenopeltis,* a trilobite in which the tips of the
thoracic segments, like the genal spines, are enormously
elongated. This specimen is from the Ordovician shales of
Wales; similar specimens can be found in France, Spain, the
Czech Republic and Morocco. These occurrences help to
define the Ordovician continent of Gondwana.

28. OPPOSITE BOTTOM *Pagetia,* a diminutive, flea-sized
trilobite with two thoracic segments and a long pygidium
the same size as the head. This trilobite is related to the blind
agnostids, but actually retains a small eye set far out on the
cheeks. This example is from the Cambrian rocks of British
Columbia, Canada.

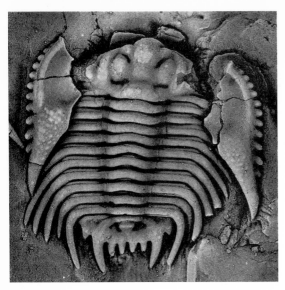

30. A beautiful moulted, cast-off exoskeleton of *Leonaspis*. See how the free cheeks have been cast off to either side as the facial sutures have been opened under the influence of moulting hormone. The "soft-shell" trilobite crawled off forwards and away to grow a new hard shell. Specimen about 1.7 cm long.

31. *Sao hirsuta* from the Cambrian of Bohemia (modern Czech Republic). The growth from babies of this species is illustrated on p. 225. This trilobite has a strongly furrowed glabella, moderate-sized eyes and tubercles on its cheeks. Sixteen thoracic segments, and the pygidium is small.

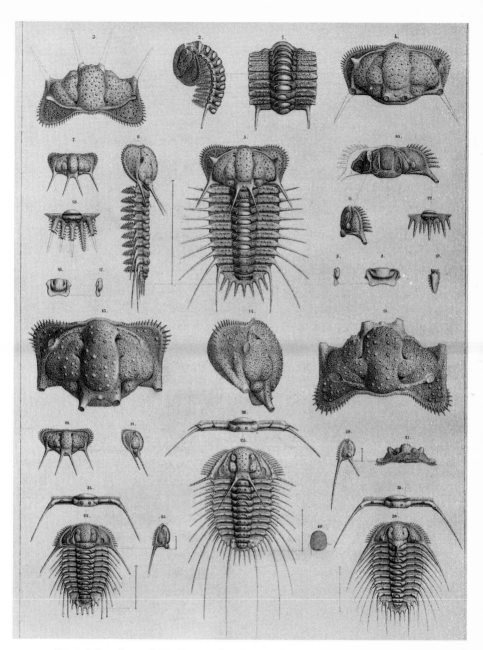

32. One of the plates from Barrande's magnificent work on the trilobites of Bohemia. The original is quarto, and so the details are even clearer. These are odontopleurid trilobites.

33. Trilobites as sympathetic magic. A piece of one of the "swallow stones" from the Cambrian of Shandong, China, abounds with heads and tails of at least three species of trilobites. (Photograph courtesy Adrian Rushton.)

34. The fantastically spiny trilobite *Comura*, from the Devonian of Morocco. The vertical spines are a masterpiece of preparation from inside the limestones in which this trilobite is found.

the earlier one, and quite right, too: scientists are supposed to move with the times. But what was lacking was any acknowledgement that the earlier version had existed at all. It was an extraordinary revision of history in favour of the present. So the root cause of Simon's explosion was not envy of Gould, but resentment of the hold he had on the past. The casual reader of *The Crucible of Creation*, unaware of the history, would never gather that the author's views had once been close to (if not actually shared with) Gould's.* Such a reader would never guess that Simon received the Schuchert Medal of the Paleontological Society of America, a signal honour, in 1991, and endorsed by Gould. "History is bunk!" Henry Ford said in 1919. Such sentiments may not be inappropriate for an automobile manufacturer, but they do little credit to a historian.

As for the trilobites, they have witnessed it all, and I shall try to take the long view through their crystal eyes, indifferent as they are to the splenetic explosions of mere humans. In the minds of their devotees they have travelled from being unexplained mysteries to becoming cousins of crustaceans; they have journeyed briefly towards being a phylum all of their own; now they have arrived back, where they belong, among the other arthropods, and closer to *Limulus* than Darwin believed. They have had theories about their closest relatives exploded and they have been caught in explosions. Maybe it is time to pack away the dynamite and let the explosion metaphor rest for a while. It's caused quite enough trouble.

---

*Some of those, like Richard Dawkins, who have responded positively to Conway Morris's criticisms of Gould also seem to have been poorly versed in the history of the "explosive" opinions. Opponents of Gould in other arenas, they have used the book as a stick to beat "the sage of Cambridge (Mass.)," operating on the principle: "my enemy's enemy is my friend."

VI

Museum

When indolent holiday crowds saunter through the galleries
past mounted skeletons of extinct animals, or scan simulacra
of dinosaurs jerkily attempting to persuade the viewer that a
hundred million years can be wished away with latex and
mechanical bones, perhaps one in a dozen of the visitors
might notice a door in the wall behind the monsters. A well-
polished mahogany entrance, it can only be opened with a
special key. Once in a while, a curator will emerge from the
door and pause for a second, as if slightly overwhelmed by
the sight of the throng. This is the door which leads away
from the show of exhibition and into another world: the reality of collections of bones and shells.

I went through that inner door for the first time more than
thirty years ago. When I joined the staff of the Natural History
Museum in London it was known in the trade as "The BM."
The British Museum: it was a grand title inherited from grand
days. The natural history collections had long since split off
from the antiquities which stock the shelves in the great building at Bloomsbury: pharaohs and pharmaceutical phials, long-
boat treasures and lorgnettes; and their research departments
of Antiquities, Egyptian, Classical, Oriental or whatever. But
the BM we remained—officially, the British Museum (Natural

History). In Italy, my colleagues still refer to us as "Il Britannico," a wonderfully absolute description that embodies an essentialist view of the nation in its collections. It was, I realized, akin to entering a holy order, complete with the vow of poverty. But I was a fortunate man, one of the few whose dreams of employment coincided exactly with reality. I, who had fallen in love with trilobites at the age of fourteen, was to be one of the few people in the world paid to do what I would have done for nothing! I was issued with The Keys. These were a set of heavy steel keys, of the kind usually used to lock up prison cells. They were held on a steel ring, and, so I was told, I had to keep them on my person at all times. On the keys were etched the words "20 shillings reward if found," dating from the days when a quid would take you and your sweetheart out for a fish supper, and the change would pay the bus home. Almost all doors opened effortlessly under their charms. There was a full-time locksmith closeted in a room that Charles Dickens would have recognized, whose job was to ensure that keys glided into locks with the intimacy of warm handshake.

I was assigned to the Palaeontology Department—to the vanished world of extinct life. When I first joined the staff of the Natural History Museum my office was part of the maze. Tucked inconspicuously under the great ceremonial Museum entrance, a Gothick cathedral door decked with motifs from nature, my office also housed most of the trilobite collections in magnificent old cabinets—the room exuded the scent of scholarship. There was even an iron balcony that ran around the midriff of the room, with more cabinets above. Outside the office, an elephant, no longer required for display, peered out from under a dust sheet. In this place the world authority on barnacles, T. H. Withers, had once worked. So had my predecessor, W. T. (Bill) Dean, who had also studied trilobites in the same room, before he had been tempted to a job in Canada. This was fortunate for me, because jobs at "The BM" were rare opportunities. A round hole opened, and a round peg was available to fill it.

A tray of trilobites from the huge collection in the Natural History Museum, London. Their labels recall essential information about where and when collected, and by whom—an archive for civilization.

When I received my first job description it said "to pursue research upon the trilobites" which was rather like saying "amuse yourself for money." To my fellow commuters on the 8:02 from Henley-on-Thames, Oxon, it probably still seems like that. As they prepare to wrestle with takeover bids, draft complex memoranda for Civil Service Committees, or design new ways to advertise beefburgers, I still march off to the trilobites. "What do you actually *do*?" they ask with genuine and bemused curiosity. Well, the basic job in a national natural history museum is to do research on species. Other things flow from it, but an understanding of diversity underpins just about everything else. I am one of a few researchers privileged to name species (in the somewhat pompous language of the trade, "a species new to science"). These are, if you like, the atoms of all subsequent speculations. This is not the glamour end of science, where galaxies are playthings and sub-atomic particles the stock in trade. This is the biological shop floor. Let me explain.

Nobody knows exactly how many living species there are. Some kinds of animals—birds, for example—are large and showy enough for new discoveries of unnamed kinds to be rather rare. But as for beetles, only a fraction of the species that thrive in trees or under rotten logs have been named: the job of nomenclature is endless (ask any Beetle Man). For the geological past the problem is a little different. We can sample only a fragment of everything that once lived. We depend on the preservation of the fossils in the rocks, itself a capricious business; we depend on luck in discovery—the right hammer in the right place at the right time. Trilobites, it will be recalled, are usually fragmentary, so we depend most of all on the persistent collector to find all the bits and pieces. Then we can set about deciding if there is a new species under the microscope. This is not an easy business.

In the first place: what is a species? Among living animals it is usually easy enough to discriminate species: closely related species differ consistently in details which can be read-ily recognized by the trained eye. Two common European birds included in the same family, song thrush and blackbird, are easily distinguished by their plumage, eggs, songs and behaviour, despite a general similarity; nor do the even more closely related mistle thrush and song thrush long confuse a practised bird watcher: the differences in their songs and habits are discriminants enough. But for fossil trilobites all we have to go on are shed carapaces. Fortunately, trilobites are rather like thrushes in one respect—they have different "plumages"—their surfaces often carry beautiful and charac-teristic details of design and sculpture that are very probably the reflection of true differences between species. Separate but related species often advertise their distinctiveness in just this way: it is a method of making sure that breeding with the right mate occurs. It is broadly the same principle that ensures that rockers bond with other rockers (studded leather jackets), rather than with, say, followers of the Hare Krishna sect (robes and shaven heads). Given well-preserved material we can recognize a fossil species as truly distinctive with almost the same confidence as with a living species. How, then, do we record this realization—turn our recognition of a new species into an official statement?

This is where scientific publication comes into the procedure. You cannot just get out of bed on a wet Monday morning and decide to make some new species. A species does not officially exist until its publication in a scientific journal. The author—often an authority—proposes the species as new and says exactly why, with appropriate illustrations. It is a serious business. You must discriminate the new species from all the others described within the same genus: in the jargon, you are obliged to "diagnose" it. This means that you have to sift through a dozen or more scientific papers, to compare the specimens in hand with all the other related species which have ever been named. This can be a laborious process, not least because the papers in question may well be published in obscure journals originating from Novosibirsk, Norwich or New Delhi. It will be obvious that to have a good library to hand is a tremendous boon to the specialist. The reference libraries attached to the great museums complement the collections as fuel does a motor. If, by mischance or laziness, you do *not* do your literature search thoroughly you could neglect a publication which actually named your species first: then, sadly, your name would be doomed to synonymy (which is a taxonomist's way of saying sunk into oblivion) because the oldest name carries priority. Scientific names are not like the street names in East European cities that change according to the political complexion of the day. They are well-nigh permanent. A rose by any other name will always be *Rosa* to the botanist.

A new species has to have a new second name—the specific name. Long years of tradition (shortly to be brought to an end) have set up rules about the classical form of species name. It has to be derived from the Greek or Latin root of the appropriate word, so, for example, a beautiful species could be christened *pulcher,* or even *pulcherrima* if it was very beautiful indeed (from the Latin). It could not be *verypretti,* or *jolliattractivi* (from the vernacular). *Rosa pulcherrima* would be quite in order. *Rosa pulcherrimus* would not, because the endings of the genus and species are supposed to agree in gender—it is a matter of euphonious sound, if nothing else. I have always rather liked this adherence to classical roots, if only because it serves to link me with the pioneer taxonomists of the eighteenth century, who wrote in Latin and probably thought in it. This much I share with the great John Ray and the incomparable Carolus Linnaeus (or Karl von Linné, to delatinize him). We are all linked through the great endeavour of classifying the natural world; across more than two hundred years we share the same passion for ordering our knowledge. I actually rather relish trawling through heavy old dictionaries compiled by learned classicists (I have Lewis and Short's *Latin Dictionary* in front of me as I write) to look up the word for, say, "blushing," or "warty," to attach it to a species, and I love to read the quotes from Ovid that justify the usage. This adherence to a past classical culture is a bond, not bondage.

The next stage is that you are obliged to fix your new scientific tag on a particular specimen, the *fons et origo* of the name, which will carry its imprimatur for ever. This is the type specimen (or holotype) of the new species. Here the museum acquires its peculiar importance. The type specimens of species are housed there in perpetuity. The collections are the ultimate reference for the variety of the natural world, past and present. Alongside the type specimens are all the other collections made from everywhere from Antarctica to Ecuador, Tien Shan or Timbuctu, an inventory of everything alive or dead. In the Natural History Museum the fossil collections alone occupy an area larger than a football pitch—and there are four floors of them. Each floor has row upon row of cabinets, and within each cabinet there may be forty drawers or so. Fifty specimens or more may reside within a single drawer: the mind soon reels if it tries to compute the number of specimens altogether in the collections. If I wish to compare a trilobite with some arthropod that is still alive I have to go to the Zoology Department. In the Spirit Building there are thousands upon thousands of jars containing fish or snake, octopus or lobster, pickled to the life. There are lizards collected by Charles Darwin. There are worms dredged from the bottom of the deep sea. Here is the one I wanted: a large relative of the woodlouse called *Serolis* which lives on the sea-floor under the Antarctic icecap. It looks superficially like a trilobite (although it is not a close relative) and I wanted to check a detail of its thoracic structure. Passing on, I do not have to be overtly anthropocentric to see in the turned-down lips of codfishes a depressed commentary on spending a hundred years in a jar. Colour fades, so that the ghostly quality of spirit preservation seems to match the antiquity of the specimens. As you slide the doors back upon this pallid parade of containers and bottles your voice automatically loses decibels. You reflect: mortality, this is your sad face; you defy decay only as a ghostly pickle.

Thus, after a species has been named, other scholars can always refer to the type specimen itself if they wish to know whether an example they have in hand is the same, or a different species. A number, usually written on a little label and glued to the specimen, is assigned by the curator, official scribe to biodiversity, which uniquely identifies that particular individual for reference (computers have made all this information much more easily available). The holotype has somewhat declined in importance since a less essentialist view of species has prevailed; it has been realized that a population of a type collection is preferable so as to give some account of variation in nature—after all, no two animals or plants are exactly alike. This enhances the importance of the *whole* collection made along with the type (some of these specimens are referred to as "paratypes"—literally, by the side of the type). In the Spirit Building there are types of species so rare that one of those pale faces looking out at me from their jars might be the only specimen known. Perhaps it is no wonder that it has a gloomy demeanour.

I look forward to the time when images of these type specimens can be summoned up globally on the World Wide Web. Suppose a researcher in Sibumasu wonders whether he has the same butterfly species as one named a hundred years before by one of the early western explorers: all he need do is log into the appropriate website on his field computer, and there will be a gallery of holotypes in living colour for him to compare with the specimen he has in his hand. All that century-long tending with naphtha balls, and the curator's dedication to numbers and records will have been justified in that moment. Only by such definitive reference can we truly know what lives where, and in what numbers. I believe it will remain necessary to summon such living, visual images for a long while yet. DNA "fingerprints" of species are becoming increasingly important, but they do not substitute for the wonderful subtlety of the human eye in judging similarities and differences. It will continue to be more practical and speedy (and cheaper) to "do it by eye." After all, fine discrimination is probably the reason why the eye and brain have become so supremely gifted in our own species.

My part in this endeavour is to be one of the few people privileged to name new species of trilobites. The routine with fossils differs little from the procedure with butterflies, although the holotypes of new species are usually less fragile than lepidopterans—I have collected many of them myself, and with a hammer. Some fossil species are rare because they are difficult to collect, which may not reflect their original rarity in nature. They may be very spiny, for example, or thin-shelled. Over the years I have named more than 150 new trilobite species, and it still gives me a little buzz to know that I have discovered a species "new to science." There have been a few genera, too. Only once have I skirted nomenclatural disaster. I decided to name a pretty new trilobite after an obscure Phrygian nymph, *Oenone,* a name I had trawled from one of my classical sources. It just sounded rather attractive, suitable for the animal. Fortunately, I discovered at the last minute that the same name had already been used for a worm, of all things. This is completely against the rule-book, which is a tome published in English and French called *Rules of Zoological Nomenclature.* I have to say that of all forms of bedtime reading, with the possible exception of Kennedy's *Latin Primer,* the *Rules* take the biscuit for being the dullest conceivable. It is a set of "thou shalts" and "thou shalt nots" for the naming of animals. Like annual accounts and railway timetables the *Rules* are necessary for the smooth running of the (naming) system,* but are also a pedant's paradise. One of the most important rules is not using the same generic name twice. Happily, I was able to quickly alter my name to *Oenonella* before it got published, and this name had never been used before—so *Oenonella* it became, and remains to this day.

When you name animals you are not allowed to be insulting to anyone, but the *Rules* do permit you to be nice and name animals after colleagues. Two Czech palaeontologists named a trilobite *Forteyops,* and there is a *Whittingtonia,* and a *Walcottaspis;* thus may the worker be commemorated in the beast. Taxonomic legend has it that somewhere in the animal kingdom there is a suffix *-chisme* (from the Greek, and pronounced "kiss me") which invites a researcher to add the names of would-be girlfriends before it—as in *Polychisme, Anachisme,* etc. I named a trilobite with a singularly hourglass-shaped glabella *monroeae* (after Marilyn), and a friend of mine named a hunchback-looking fossil *quasimodo.* These little diversions actually help to make names more memorable.

---

*To remind readers who may not be familiar with taxonomy, the generic name is the first one, and capitalized, and a given genus may contain a number of species, characterized individually by the second, specific name, which is not capitalized. Scientific names are always italicized to distinguish them from the vernacular.

The *Rules* do not allow you to name a species after yourself, but jokes in naming are permitted if they do not cause affront. It is not flattering to name a new species *jonesi* in honour of Jones, if you go on to describe it as "this diminutive and undistinctive species is a typical inhabitant of dungheaps." Usually, species names just tell you, in Latin or Greek, something about the animal in question: *Agnostus pisiformis* (the pea-like agnostid trilobite), *Paradoxides oelandicus* (the Paradoxides from the island of Oland), and so on.

To the name is appended the namer. Thus, an unusually attractive Ordovician trilobite from Spitsbergen (named after my wife, naturally) is correctly known as *Parapilekia jacquelinae* Fortey, 1980. This detail serves the useful purpose of directing subsequent researchers to the reference where the species was originally described and named: a paper published by Fortey in 1980. In the case of species named a century ago, or more, subsequent accounts of the same species (revisions) may also have been prepared. Many palaeontologists that I have never met face to face probably know me as an appendix to a name. I hope that they will be astounded by my youthfulness when we finally get to meet.

The familiar, if slightly garbled, quote from Romeo and Juliet, "A rose by any other name would smell as sweet" implies that the naming of names serves little purpose. This same stricture might be thought to apply to the kind of science which was famously labelled "stamp collecting" by the physicist Ernest Rutherford—and taxonomy may well have been in his mind. That view could not be more misguided. Although the dubbing of scientific names may be fun, these same names can also be deployed for real intellectual purposes. Critical identification is central to some of the important questions I shall examine below. How can you talk about the diversity of life in the past unless the units of measurement (species, genera and the like) are accurately defined by competent taxonomists? How can you speculate on evolution unless you know that the species you are examining are likely to be real entities?

How can you cerebrate about the ancient geography of life if there are no reliable labels to place on this animal occurring in this continent and that one on the other? Three rhetorical questions in a row is about as much of an indulgence as any book should be allowed, so I shall merely answer my own challenges with: "Of course you can't" and get off my soap box.

But I should say, *pace* Rutherford, that the perfectly amiable activity of stamp collecting differs from scientific taxonomy. For any postage stamp, we can look up the date of issue in Stanley Gibbons's catalogue, check the colour, check the watermark, and check the perforation, even check the current valuation—there is a single, unique, *right* answer which identifies any stamp. But all questions in real science are journeys towards the right answer. It is appropriate to recall Robert Louis Stevenson's aphorism: "to travel hopefully is a better thing than to arrive." Science exists in a continuous spirit of moving optimism. We can never know *for sure* that a trilobite species recognized by my considered and experienced observations of features of the glabella and pygidium was actually a real, biological species when it lived hundreds of millions of years ago. It often happens that another worker comes along and disagrees with my species, alleging that it might merely be a variety (usually of one of his). There is no final arbiter on such matters. Nor can we ever reconstruct a long-vanished biological world with certainty, for every reconstruction is only as good as the scientific inferences that have been made about it, and these inferences are subject to continuous change. Here are two examples. First, it is only a few years since we realised that there were phases of high and low carbon dioxide atmospheres in the past—producing "greenhouse" and "icehouse" worlds, respectively. These conditions influence almost everything on the Earth's surface, from sediment type to sunlight—and must affect living organisms. Second, at one time it was believed that fishes only began to evolve at the end of the Silurian, but now new discoveries have shown that there were primitive relatives of fish alongside trilobites for most of their history; this in turn forces us to look at the ecology of the Ordovician with new eyes. These are changes in perception of the past. Even as time's arrow moves forwards, the past is redesigned in retrospect.

In the nineteenth century, a museum sprang up in almost every large town of the developed world. This was partly the consequence of a widespread belief in their improving value, in both an educational and moral sense. It was often a matter of civic pride. Whereas in medieval times wealthy wool merchants endowed churches, their equivalents in the industrial age endowed museums. In Britain there are museums in Hardy country, in Dorchester and Lyme Regis; and in Wordsworth country, the Lake District, as at Keswick; and of course in the great industrial cities: Manchester, Liverpool, Birmingham and Leeds. In the United States every major city in the East has a museum, some of them associated with great philanthropic names like Peabody (Yale) or Carnegie (Pittsburgh). You can find similar museums in Australia and central Europe. Many of these museums have natural history collections, as well as the spoils of the founder's taste in art. Often, their collections include important type specimens. For the researcher, tracing these specimens can be an adventure, because not every small museum knows exactly what it has got. My friend Adrian Rushton discovered some specimens of trilobites from the Keswick Museum described by J. Postlethwaite in a book on *Mines and Minerals of the Lake District*, published in a very limited edition in the 1880s. Arcane information, you might suppose, until you learn that trilobites are inordinately rare in the Lake District, and that Mr. Postlethwaite named and found many of them. Then it should be added that the whole history of the Lake District in its geological youth hinges on the identity of these rare animals.

The creation of great museums is one of the hallmarks of civilization. During periods of cultural decline such treasuries of knowledge are abandoned—witness the virtual loss of great works of Greek science during the Dark Ages. They

The horseshoe "crab" *Limulus,* now regarded as the closest living relative of the trilobites. (Photograph courtesy Richard Kolar, Oxford Scientific Films.)

were saved because the caliph al-Mamun ordered the construction of a museum and library in Baghdad, the Bait al-Hikma (House of Wisdom), completed in 833. This was no dull storehouse—it was a vital link between classical civilization and the Renaissance. I see the natural history museums of today as bearing witness to what mankind will do to his planet and the creatures he shares it with. Even the most curious items may yet prove their worth. Consider the collection of dog breeds that Lord Rothschild made, now stored outside London at Tring in the very model of a nineteenth-century parade-ground exhibition. Surely this kind of thing is passé and redundant? But is it not possible that a future researcher might want to investigate the history of domestication, and that the old skins might be the source of molecular information? Every dog shall have his DNA; and a great museum should never die.

VII

A Matter of Life and Death

Trilobites, like all other creatures, evolved. I don't just mean that they changed through time: that much is obvious. Trilobites like *Olenellus*, from the Lower Cambrian, are different from those in the late Cambrian, and these in turn are distinct from those in the Ordovician, which differ again from specimens found in the overlying Silurian and Devonian strata. With even a little expertise a trilobite lover will be able to cast an eye over a group of fossils and guess their age, even if they cannot put an exact name to them. They respond to what bird watchers call the "jizz," a kind of overall impression that rarely misleads. Clearly, trilobites replaced one another through the geological ages. In most rock sections, though, we assume that every trilobite novelty that appears was an evolutionary innovation even when the rocks themselves may often provide no details of its origin. To see "evolution in action" is rather rare. This rather mundane truth has been misappropriated by creation "scientists" as evidence that "fossils don't provide support for evolution"—which is not the same thing at all. In fact, the order of appearance of trilobites is certainly *consistent* with evolution: Cambrian trilobites have more primitive characteristics than those in the Ordovician and younger, as we have already seen in the case of the peculiar, and evolutionarily advanced, schizochroal eye. It is just that it is genuinely difficult to catch the appearance of new species in the act of creation. In a burglary, it is rare to come upon the scene with the miscreant standing there, caught red-handed and carrying the swag: the subsequent mayhem is what people usually come home to. So with the generation of species—the endurance of the species after the event is relatively long, so that this part of their history will be more likely to be discovered, just by the laws of probability, which have no respect for wishful thinking whether creationist or Darwinian. To maintain the criminological metaphor: sampling bias alone leaves little chance of conviction.

So examples where you *can* see evolution at work become doubly precious. Our own genus, *Homo,* and its several species related to modern Person, is not a particularly good subject: hominids have few fossils and the most arguments. This does not mean that we are not finding out more and more about hominids—new fossils turn up every year—it is just that human history may not be the best choice for studying features of speciation itself. The fossils are still too rare. By contrast, trilobite examples have been central to some of the most vigorous debates about how evolution happened. Since they are complex and abundant fossils, they might be expected to be particularly useful "experimental material" to cast light upon how new species are generated. In the laboratory, another arthropod, the fruit fly *Drosophila,* has for many years been used as the experimental animal for genetics-in-action. The classical studies on inheritance were carried out upon this little fly. When the roles of specific genes were investigated—most recently the family of HOX genes that control sequence in development—it was *Drosophila* that was manipulated to produce hopeless but informative monsters with extra pairs of wings, or legs in place of antennae. Fossils of flies are too delicate for any but the most exceptional preservation in amber. Maybe robust trilobites could be the fruit flies of the rocks.

What is required as a test case is a sequence of species found one after another, in the same rock succession, which can reasonably be interpreted as having an ancestor–descendant relationship. There should be large numbers of specimens collectable from many rock layers recording the whole history of both the older and younger species—so that measurements can be made on what happens to the shape of the animals throughout the time of deposition of the rocks, not least to convince the sceptics who doubt that any evolution is going on at all. Unusual sedimentary successions are required which lack prolonged breaks in deposition—for hiatuses may disguise the very moment of evolution into a different species. Most rock sequences are, in practice, incomplete. It is not altogether surprising that these critical initial conditions are rarely met: most rock successions fail on one criterion or another. The most suitable deposits—and these are of comparatively young geological age—are the rocks which accumulated on deep-sea floors, where a continuous rain of plankton ticks off time in a settling mist of tiny shells. These little fossils, often belonging to single-celled, calcite-shelled organisms called foraminiferans, have provided many of the best evolutionary case histories, not least because they are so abundant. A handful of rock may yield hundreds of specimens. But their small size also means that they tend to be rather simple—a few bubble-like chambers a millimetre across. And maybe plankton has evolutionary properties different from their bottom-living relatives. Trilobites might, after all, provide an example which is much more typical of most marine life. The immediate problem is the obligation to collect samples large enough for a convincing study. This means hours of bashing rocks, even if the fossils are quite common. You cannot study trilobites as you might gas a few generations of fruit flies to see the changes in the genealogy. The long-term application of brute force is required to get a decent sample. There have been several scientists who have this kind of doggedness, strength and patience. As we shall see, they came to quite opposite conclusions on what the trilobites had to say about the origin of new species.

The phrase "punctuated equilibrium" has become rather familiar: I recently heard it referred to horribly as "punk eck" by an Australian philosopher of science. Few laymen, or even scientists, realize that its genesis was thoroughly grounded in trilobites. In the late 1960s a young American, Niles Eldredge, was studying the Devonian trilobite genus *Phacops* in North America. We have already met *Phacops* as the possessor of marvellously complex schizochroal eyes, in which each lens was a tiny calcite spheroid, separated by a little inter-lensar sclera. The number of lenses is relatively few, and they are easily countable under a microscope. It is a very common fossil in the appropriate strata exposed in the states of New York, Iowa and Oklahoma, and many other localities besides. It is often well-preserved in limestones, a preservation that permits its most intimate details to be inspected. A few taps in the right locality, and out a *Phacops* will pop (usually the cephalon) as if to say, "jeepers, creepers! how about *my* peepers?" *Phacops* provides one of those rare cases where there is a hope of discovering the particulars of the evolutionary process between one species and the next, so prolific is the fossil record. To his credit, Niles realized this scientific opportunity quite early in his studies.

Niles noticed changes in the arrangement of the lenses in the eyes of *Phacops* species. He counted the lenses in their "dorso-ventral files"—the number of lenses in each row counted from the top to the bottom of the eyes. Let him describe in his own words the observations that he confirmed while writing his dissertation, as recalled recently in his book *The Pattern of Evolution:*

Bingo! Another pattern leapt out: Populations in the Appalachian basin seemed to be invariably 17 dorsoventral files, for almost the entire length of Middle Devonian time . . . In the Midwest the story was altogether different: For 2 million years or so, the number remained stable at 18; thereafter, for at least another 2 million years, the number was also stable, but it was 17, not 18; for the very last part of my time frame, the trilobites had 15 dorsoventral files . . . large chunks of time were missing in the midwestern rock sections precisely when the changeover from 18 to 17, and from 17 to 15 files occurred. The seas in which *Phacops rana* . . . lived simply dried up during those intervals . . . the change from 18 to 17, and later from 17 to 15 dorsoventral files in the Midwest simply marked a repopulating of the seas. The pattern suggests that the 18 file form became extinct when the midwestern seas first withdrew; when the seas returned the 17 file form was available and took over the newly reconstituted marine habitat.

Niles focused especially on the changes in the eyes because he saw that here were the crucial characters for defining the species. If he had been studying birds it might have been tail feathers or song. If he had been studying molluscs it might have been the patterns on the shell. Each animal parades its own peculiarities to establish its own identity. Species flaunt their personalities to signal to their own kind.

Two conclusions come out of Niles's observations on Devonian *Phacops*. The first is that the events portraying generation of new species are rather hard to observe—they seem always to happen "somewhere else." However, subsequent to the appearance of a new species, the successful innovation often invades, and replaces, an earlier species. In some cases Niles knew where a new species originated, but entirely transitional populations between new and old species were hard to find. This phenomenon may be compared with the Beatles overtaking the pop music scene in the sixties—only to be replaced by Bee Gees in the seventies, or Michael Jackson in the eighties. The early discs are the collector's items, the later ones characterize a whole cultural phase and are as common as unsuccessful lottery tickets. Thus, the new species often starts out as a relatively small population somewhere on the edge of the range of the (then) dominant species; geographical separation is what produces a species difference. But when its time comes the new species replaces the ancestor and enjoys its own hours of glory. This is where the Harvard University biologist Ernst Mayr's influence was crucial, for Mayr had already observed that in the living world new species often seemed to have been generated as a result of geographical isolation of a group of individuals (this process he termed allopatry); disjunct populations may have acted as the "motor" for evolutionary change. The isolated group had its gene flow with the mother species interrupted; and separation alone could breed novelty. Evolution did, really, happen "somewhere else."

Niles's second conclusion was that once a new species had arrived it endured, often for a long time, with little change. We may not see the origin of the species, but we do see its acme. Like the burglar that broke in at the dead of night, the significant event was clandestine: we see the aftermath, but not the deed. The mundane demonstration of this in the rocks is the observation that a particular *Phacops* species, once it has appeared, endured for a long time with very little change. To the practical field geologist this means that much breakage of rock to recover trilobite fossils by working through metre after metre of strata—the bloodied fingers, the wet feet, the mosquito bites (especially in New York)—is rewarded by a cry of "no change!" This is hard, hard labour to show the absence of something, which in some scientific circles is often called negative evidence; strenuous work for no result, you might suppose.

Except that the result was highly important. Species, said Niles, originate allopatrically—"somewhere else." When one of these species successfully invades, and then replaces, its ancestor it endures for a considerable time. Life proceeds by fits and starts; a species lasts until it is replaced by another—and that replacement is rapid. Taken together, the two ideas—the endurance of species, and allopatric speciation—make up the conceptual basis for punctuated equilibrium; the choice of words for this theory will by now be obvious. The equilibrium is the enduring phase of a species' life; the punctuation is its sudden replacement. *We shall all be changed, in the twinkling of an eye,* as Corinthians puts it. The new theory was set up in opposition to the notion of "gradualism," a slow and more or less continuous change or shift that nudged whole populations towards the new species. This was considered to be the dominant model for evolution in the aftermath of the "modern synthesis" of evolution in the 1930s—and a rather supine acceptance of this creed ensured that the "punctuated" view, when it appeared, was heralded as startlingly novel. Niles joined forces with Steve Gould to present the new model, and with considerable success. Their original 1971 paper achieved an enormous "citation index"—this is a measure of the influence of a piece of published work by the number of times it is quoted in the bibliography by other workers. The punctuational description of evolutionary change lent itself readily to metaphor, and other observers were quick to point out similarities in several areas of science and culture that were not just concerned with speciation of animals. Even human history could, with a little massaging, be described in terms which seemed consistent with "punk eck": for example, cultural revolution was often followed by dynasties which finished in a state of stasis. Gibbon's *Decline and Fall of the Roman Empire* might exemplify the inevitability of historical patterns as much as human foibles. In a book that appeared some years after the seminal paper, *Time Frames,* Niles himself went along with the pervasiveness of punctations in history. The story of our planet's development, it seemed by then, was a tale told mostly by jerks.

This perceptual revolution was perhaps an inordinate burden to place upon the cephalon of the humble, if pretty, trilobite *Phacops*, whose eyes alone would have been able to see the evolutionary truth. As we have already discovered, they probably saw rather acutely. Other punctuational examples from the fossil record soon joined the "ay" voters. Quite soon, the punctuation explanation provided a rational counter to creationists who sought to exaggerate the rarity of "missing links" in the fossil record to counter evolutionary theory. On the contrary, such gaps might be just what evolution *demanded*. As a life-long rationalist in defence of the explicable against the numinous, Steve Gould welcomed this arsenal of ammunition in his campaign to educate those who denied the magnificent narrative of Earth history in favour of a week's labour by the Creator. *Phacops* had become tied up with some contentious company. Arguments raged between bible purists and evolutionists: it was a case of trilobite-by-jury.

Niles was not the first to make the observation of "punctuated" change in trilobites. Nearly forty years earlier, a German from the University of Greifswald, Rudolf Kaufmann, had drawn similar conclusions from a minute study of late Cambrian olenid trilobites from the Alum shale of Scandinavia. We have already met olenid trilobites of the genus *Triarthrus*; this was one of the trilobites of which the legs and antennae were first known in detail. It will be recalled that olenids lived in a special environment in which the sea-floor was low in oxygen, while in the sediment below there was a complete lack of oxygen and a high concentration of sulphur. I even suggested that the olenids may have cultivated colourless sulphur bacteria as symbionts. About 500 million years ago in the late Cambrian an Olenid Sea spread across the whole of southern Scandinavia, an inundation that persisted for something like fifteen million years. This was a special time, because during much of this long period there was nearly continuous deposition of dark, shaly strata, which now yield frequent trilobite fossils. If you can find a quarry exposing the Alum shales, you break up the smelly nodules known as "stinkstones"—often about the size of a rugby football—and find beautiful and abundant

A portrait of Rudolf Kaufmann, the tragic German trilobite palaeontologist.

trilobite remains. The Alum shale is a famous example of a "condensed deposit" where much geological time is crammed into a thin sequence of strata that accumulated without major breaks. This approaches the ideal case for carrying out evolutionary "experiments" in the field. Kaufmann was astute enough to recognize this, and made careful collections from successive layers of strata to observe the subtlest of changes through time. Niles did fully acknowledge this pioneering work, published in 1933, work which would certainly have been more widely known had it not been published in a journal of Greifswald University with very limited circulation. (This recalls Gregor Mendel's crucial experiments on inheritance in plants carried out in the Czech town of Brno, and their long struggle into the light of international science; it might be even worse today, with ten times more journals competing for attention.)

What Kaufmann observed was that several species of *Olenus* (see p. 70) appeared suddenly in the rock sections, and then had comparatively long ranges. But during their "lifetime" the species were not static; instead they showed small variations, especially in the shape of the pygidia, which became progressively narrower and longer through time. The same changes happened to the pygidia of different *Olenus* species. Kaufmann clearly showed the invasion of a species from elsewhere into the Olenid Sea of Scandinavia, thus presenting a graphic illustration of allopatry before it was a recognized concept. Furthermore, he based his results on large collections, and analysed the results in a quantitative fashion. Euan Clarkson has revisited the famous Swedish quarry at Andrarum in the last few years, and repeated Kaufmann's observations. Clearly, this was a remarkable and far-seeing scientist.

I was puzzled by Rudolf Kaufmann's apparent disappearance from the trilobite firmament after this seminal paper. Scientists can usually be mapped through a career of twenty-five years or more (in some cases you wish it were not quite so long). They leave behind a legacy of papers, which can be used to track an intellectual lifetime—literally a paper trail; not least, people tend to quote themselves, so that you can find out a biography from a bibliography at the end of a paper. To an experienced researcher with a good museum library such sleuthing is almost routine. Not so with Rudolf Kaufmann; he simply vanished. It was not until 1998 that I discovered why. It is an extraordinary and moving story.

That we know the story at all is because Reinhard Kaiser bought a mixed bundle of letters and postcards at a stamp auction in Frankfurt-am-Main in 1991. He paid 500 deutsche marks for the job lot. Among the batch were Rudolf's letters to Ingeborg Magnusson, his Swedish lover. Kaiser was so engaged by the poignancy of the tale they revealed that he discovered who Rudolf Kaufmann was, and pieced together the narrative. Sadly, Ingeborg's letters to Rudolf have not survived. Her devotion is indicated by the fact that she never married, and kept the letters from Rudolf until her death in 1972. They had met in 1935 when he went to Bologna, the ancient university city in north-eastern Italy; he had fallen instantly in love with the dark-haired Swedish girl. She was only united with him again for a few days between their Bologna idyll and his tragic death. The story is glimpsed in fragments through their correspondence; it tells of his attempts to reach her Swedish haven during the ghastly years of Hitler's dictatorship. Kaiser called his story *Königskinder* (king's children) after a comparison Kaufmann had made in one of his letters to the figures in a folksong:

> *Es waren zwei Königskinder, die hatten einander so lieb.*
> *Sie konnten zusammen nicht kommen, das Wasser war viel zu tief.*
> (There were two king's children, they loved each other.
> They could not come together because the water was too deep.)

Rudolf Kaufmann was Jewish by birth, although a practising Christian. His masterly work on olenids was published just two days after Hitler became Chancellor, and took over the government, on 30 January 1933. Kaufmann was fired from his job at Greifswald University almost immediately. This did not prevent him from pursuing his palaeontological studies outside Germany, but the Bologna visit where he met Inge was to be his last.

Kaufmann was well aware of the importance of his studies on the trilobites which were his second love. He wrote to Inge that he would send her everything that he had written as a geologist because "it will soon no longer be true that I have done all this research," a reference to Hitler's denial of the intellectual achievements of Jewry. "I am very proud of my great work on trilobites. I have been able to prove that there is a determined development in the life history of these animals. I think I will be much more famous than at present many years from now, when zoologists and palaeontologists begin fully to understand my work." He has not yet had his due.

While separated from his lover, Kaufmann succumbed to temptation. He was imprisoned in Coburg in 1936 for illegal sexual congress with an Aryan woman. In fact, he had visited a prostitute and caught a venereal infection; it was the doctor who treated him who subsequently betrayed him to the police. On 13 August 1936 he wrote to Inge: "I wanted to confess everything to you in Sweden, but now it's too late for that. I am no longer worthy of you, and I beseech you to try to forget me. I thank you for your faithful, pure love . . . you were so good to me and I proved myself to be weak, and now I have to pay for my actions . . . So much has already been taken from me in my life; my mother, my beloved career . . . This time, however, I have failed out of foolishness and I must bear your decision."

Although Inge readily forgave him, his lapse cost him dear. By the time he got out of prison on 12 October 1939, hostilities had already commenced. Had he been released six weeks earlier, before Britain and France declared war on Germany on 3 September, he might well have made good his escape. Several other trilobite experts fled from Nazism. Alexander Armin Öpik was a member of a distinguished Estonian scientific family (his brother was a famous astronomer) who eventually escaped to Australia; his fellow Estonian Valdar Jaanusson became doyen of Scandinavian trilobite experts at the Swedish National Museum. The Baltic Sea was not the same barrier to them as it was to the "King's children." By November 1939 Kaufmann was in Cologne. "And when I am alone?" he wrote. "I am with my much-loved Trilobites, of which you can already be jealous. I recently read the Odyssey. I must learn from Odysseus . . . see how he bore his longing for Penelope, and treat it as though it were written for me." He remained optimistic in the face of growing evidence that would have reduced somebody less buoyant to despair. But, gradually, his hopes dimmed of ever being united with his lover; by July 1940 he confessed that he "had no courage for the future." He doubted whether he had the strength to continue. "Yet I cannot lie. The short time we were together, and the long separation, and the worries of last month, and of every moment, and the hopelessness of the future. These are all to blame . . . Try to be free, as free as you can. There is too little hope that we will see each other for a long time in the foreseeable future. Wouldn't it be better if we weren't so close, if we did not have to torture ourselves so much? . . . I take you into my arms once again and kiss you with all my heart."

By 1941 Rudolf Kaufmann was in exile in Kaunas, Lithuania. The Baltic Sea was not far away, but it was still too deep and too wide. He had given up hope of joining Inge. He was shot in cold blood by two guards who happened to recognize him, and became just another digit in the most disgraceful statistic of the twentieth century. The progenitor of punctuation had fallen victim to one of the most aberrant cultural shifts in human history. Reinhard Kaiser discovered photographs of Rudolf: with his slicked-back black hair, handsome regular features, and a rather Germanic high seriousness, he was the very model of a young Professor, and you can understand Ingeborg's passion for him.

So it is that the study of trilobite evolution casts light both upon the fundamentals of how new species are generated, and, as I discovered through Kaiser's detective work, also upon the paradoxes of the best and the worst in the human condition. The enthusiasm Kaufmann felt for his trilobites, and his passion for the investigation of the truth they revealed, was matched by his love for Ingeborg Magnusson. Who knows what reputation he might have forged if he had been permitted to follow both his heart and his mind?

. . .

Punctuated equilibrium was not the only pattern of evolution that trilobites revealed. During the late 1970s another young man, an Englishman this time, was studying the trilobites of the area around the old spa towns of Builth and Llandrindod Wells, part of the borderland region between England and Wales. This is hilly country, a patchwork of deep green fields where sheep are the main crop, interspersed with wooded holts and little steep-sided valleys where fallen branches thickly covered with feathery mosses vie with tangled brambles to impede the progress of the geologist in his wellington boots and thorn-proof jacket. Pheasants suddenly start from the undergrowth uttering sharp cries. In the streams you will meet toads quietly going about their business among the ferns. There is a sense of moist abundance: so rich is the vegetation that the overhanging foliage cuts out most of the light. The best time to do fieldwork is in the spring, before the stinging nettles have sprouted and hidden the rocks, and before the generous leaves of chestnuts or hazel have fully unfurled. In late April there are bluebells on the banks, and drifts of yellow celandine, and blackbirds everywhere. In the banks of the streams, encrusted with liverworts, there are heavy, black mudstones, what the Welsh term "rab," which you can pick out in small slabs with the pointed end of the geological hammer. Split the slabs in the right direction and you will be rewarded with trilobites. Carefully work your way upstream to sample successive rock beds, and you will be party to a narrative that spells out evolutionary changes through geological time. The rock succession is relatively thick—many tens of metres—compared with the "condensed" shales that Rudolf Kaufmann had sampled in Sweden. This is an advantage: if you cannot find fossils in a few feet of strata the chances are that you have not passed a major geological event—in Sweden, a comparable thickness would be a vital contribution to the narrative of time. The rocks are of Ordovician age, about 470 million years old.

Peter Sheldon spent years collecting these dark rocks. With

consummate patience, he split the ungrateful shales for month after month, slowly accumulating and labelling samples of trilobites which he would later analyse. Mostly, he found isolated tails or heads; occasionally he was rewarded with a whole specimen. The commonest of the trilobites was an old friend, the asaphid trilobite *Ogygiocarella*, which will be recalled as the first ever trilobite to be described—as a "flatfish"—by Dr. Lhwyd in the vicinity of the South Wales town of Llandeilo. In these dull shales there was a trawl of flatfish sufficient to satisfy Neptune himself. The semicircular, furrowed tails split from the enclosing *rab*; with a little skill they could be displayed perfectly, little fans mostly larger than a butterfly's wing. The narrow axis occupies the central part, and is divided into numerous rings; the flat pleural fields are divided into an equal number of ribs which get shorter and rather less distinct towards the posterior of the tail. Alongside these large trilobites, and slightly less abundant, are smaller ones no more than a few centimetres long, and more commonly found in a complete state. These belong to the blind genus *Cnemidopyge* (fig. 25), a trilobite with a semicircular head and with a long spine extending forwards from the front of the glabella. This animal had only six flat thoracic segments, and a triangular pygidium which was, like that of *Ogygiocarella*, strongly furrowed. Occasionally, it is possible to find one that has rolled up. There were other trilobites, too, including a close relative of the "Dudley Bug" *Calymene*, and distinctive little medallion-like trinucleids.

All these animals were collected by the persistent Peter Sheldon. He got to know the country and the strata with an intimacy which must surpass even the farmers who own the land. He had to move from one stream to another to obtain a complete picture of the succession of rocks by carefully tracing an individual stratum across country. The work was very slow, and made still slower by the fact that Peter is one of those enthusiasts who love to explain their work to all comers. He is friendly, sempiternally youthful and tirelessly optimistic, all of which has served his dedication as a teacher at the Open University for a number of years. While he was writing the dissertation for his PhD he was always returning for "just one more collection." In the trilobite world he became notorious for his reluctance to leave the outcrop and write up. Normally PhD theses are supposed to take three years, four at most, but Peter's seemed to go on for ever. He dodged the censorious glances of the senior faculty, and just plugged on, splitting more and more black shales and collecting more and more trilobites. Just when he might have tried the patience of his supervisor beyond endurance—bingo! (as Niles Eldredge would have put it) he published the result in *Nature*. It instantly made him quite famous.

What he claimed was that the trilobites from the Ordovician strata around Builth Wells showed a kind of gradualistic change through time. He showed that this kind of change affected not only one, but several of the different trilobites that ranged through the black mudstones and shales. The most obvious example from the largest and commonest trilobite, *Ogygiocarella debuchii*, showed an increase in the number of ribs on the pygidium, from 11 to 14 on average. In the last century the pioneering British trilobite expert, John Salter, had recognized the form with more ribs as "variety *angustissima*." These are exactly the kind of subtle changes which trilobitologists use to distinguish fossil species. What Peter showed was that there was a seamless transition between *debuchii* and *angustissima*. The large populations he collected showed a good deal of variation at any one level—that is, there were specimens with various numbers of ribs found together at any one time. On some examples, there were even half-ribs on one side of the pygidium, but not the other. In general, though, there was an unmistakeable trend—at the population level— to having more ribs through geological time. When he collected at a very minute scale he found that there were even short-lived backward steps in rib number within the overall increasing trend. The progression from one form to another resembled the tottering steps of the cartoon drunk rather than a smooth progression. Even more exciting, Peter found that the tail of *Cnemidopyge* was undergoing a parallel series of changes *at the same time through the same strata—Ogygiocarella* was not unique.

There were subtler changes in some of the other trilobites, too. It all pointed to a very different mechanism for change from that which had affected *Phacops*. Even if the shales accumulated at a rapid pace under the Ordovician sea, each of these changes must have taken several million years to have proceeded to completion—this is change of a different order of magnitude from the rapid alterations induced by allopatric separation. It is actually rather difficult to think of a mechanism that would reset something this *slowly*—after all, fruit fly breeding experiments can drive an advantageous mutation throughout a population within a comparatively modest number of generations. Could it be "drift," with no particular adaptive function? Other critics suggested that the changes seen in the pygidia were not evolution at all, but were a response to slowly changing conditions on the sea-floor. These kinds of pygidial modification might be a response to changing oxygen levels, for example. Where gradualistic change had been observed in other fossil examples it was usually plankton that displayed it. There was really no question that *Ogygiocarella* and its friends were bottom-dwellers, so this example retains its puzzles and disputes. What nobody questions is the reality of what Peter observed and its relevance to evolutionary questions, and who could fail to admire the unusual persistence that inspired the observations?

There is another test case for evolution where trilobites have assumed a starring role: as a field demonstration of what is known as heterochrony. The term is Greek for "other time," and is simply explained. Trilobites grew from a larval state starting as little discs—protaspides—a millimetre or less in length. They then passed through several moults as they increased in size to achieve the adult state. The small growth stages first show the demarcation of the (proto) tail from the head. Then the thoracic segments are "released" into the thorax, one at a time in most species, and very probably at successive moults, until they reach the adult number of segments. Thereafter, in most trilobites, segment number remains the same even though the trilobite may increase quite dramatically in size—the mature number of segments in the thorax may be achieved while the trilobite is still quite small. Changes to almost all parts of the carapace occur during this growth, properly called ontogeny. The growth story is known for a large array of trilobite species, and this makes them especially important in studying the relationship between the development of the individual (ontogeny) and the appearance of novel features in new species (phylogeny).

The smallest trilobite, *Acanthopleurella*, a diminutive, blind trilobite with four thoracic segments mature at just over a millimetre in length. Ordovician, Shropshire, western England.

Some years ago Adrian Rushton and I noticed that the tiny trilobite *Acanthopleurella*, with only four thoracic segments, was probably related to *Shumardia* (p. 231) with six. *Acanthopleurella* is even smaller than *Shumardia*, and we concluded that it was derived from its ancestor with six segments by a process of "arrested development"—it became sexually mature when only four segments had been released. This explained its minute size, mature at just over a millimetre. It was particularly satisfactory that we identified *Shumardia* as an ancestor, since Sir James Stubblefield had used this very trilobite to prove that the thorax grew during ontogeny by release of thoracic segments from the front edge of the pygidium—they were "budded off" there and moved forwards, like customers in a growing queue, as more were added on behind. We could feel confident that the last two segments were repressed in *Acanthopleurella* by comparison with *Shumardia*.

At about the same time, Ken McNamara was making more detailed observations on the Lower Cambrian trilobite *Olenellus* from Scotland. *Olenellus* will be remembered as the most primitive trilobite from our parade, a form with numerous thoracic segments and a tiny pygidium. Soft, yellowish shales crop out in a few places in the bleak but beautiful coastal north-west Highlands, where sphagnum bogs and tussock grassland are populated by a few highlanders and rather more sheep, both equally hardy. This is famous ground for geology, because the interpretation of the Moine Thrust there was the subject of a great Highlands Controversy in the latter half of the nineteenth century. The Cambrian shales lie underneath the older Moine rocks, which were eventually proved to have been thrust over on top of the Cambrian. The trilobites from the shales provided undeniable evidence of their age. I have spent a very wet and cold summer field season in the area around the little town of Durness—about as far north-west as you can go on mainland Britain—snuffling rather unsuccessfully over the outcrop after fossils. My woollen socks spent most of the time drying over a meagre butane fire. I was left with redoubled admiration for the geologists Peach and Horne who had worked out every detail of this uncompromising landscape, and most of it on foot. In the century since these heroes solved the geological map we've become a namby-pamby lot.

Ken McNamara was interested in the trilobites for reasons other than their antiquity. He had realized that the genesis of several species of *Olenellus* could be understood very readily as an example of heterochrony. He already knew the development (ontogeny) of the commonest species, *Olenellus lapworthi*, named for the great Charles Lapworth, the scientist who in turn had named the Ordovician. Ken recognized that various other *Olenellus* species from Scotland had adults that resembled the *immature* growth stages of *O. lapworthi*. To cite one feature, the single pair of spines on the edge of the cephalic shield of *O. lapworthi* were positioned at the genal angle, more or less on a level with the back end of the glabella. In various of these other species the spines were shifted forwards, opposite one or another of the glabellar furrows, so that the back border of the head curved forwards to the genal angle. This is exactly what was found in *small* growth stages of *O. lapworthi*, but lost in the adult. Similar kinds of changes happened with the size and position of the eyes. Most striking of all was a small trilobite with *three* pairs of spines on the edge of the headshield, so spiky a creature that it had been dubbed *armatus* by its discoverer, and so different from *Olenellus* that it had been placed in a separate genus, *Olenelloides*. Ken McNamara realized that this oddity closely resembled a "blown up" version of one of the smallest growth stages of *Olenellus lapworthi*. Not only that, this strange little animal only had nine thoracic segments, compared with the fourteen or so of *O. lapworthi*. Looked at in this light, *O. armatus* seemed to cry out: "I'm an overgrown baby!"

Ken arranged five *Olenellus* species (with *lapworthi* at the base) in a kind of sequence of progressive babyhood, culminating in *O. armatus*. He believed that *O. lapworthi* might have lived in the deepest water, and considered that *O. armatus* probably lived in the shallowest marine environment, with the other species arranged in between. He speculated that warmer, shallower Cambrian environments stimulated earlier maturation. The five successive species then slotted neatly into different ecological niches related to water depth. However interpreted, *Olenellus* provided the most vivid demonstration of how apparently major differences between species may actually just be a question of altering *rates* of development. *O. armatus* and *O. lapworthi* look like very different trilobites—so much so that they had once been placed into different genera—but they are fundamentally related, as might be clocks with the same mechanisms but different faces. Similar heterochronic variations have now been recognized in many different kinds of animals and plants: differential development seems to be an important source of novelty throughout the biological world. *Olenellus's* ancient example gives a new twist to Wordsworth's aphoristic line "The child is father of the man."*

If the child can become a precocious grown-up, there are also counter examples, where the immature phase of a descendant species resembles its ancestor. The descendant does everything that the ancestor does during development, but adds on a little more, a novelty of its own unseen in any earlier, or more primitive species. This is the familiar case of recapitulation, of "ontogeny repeating phylogeny," which biology students used to have to learn as a kind of mantra. The grossly simplified version that once portrayed the human embryo as passing progressively through protozoan and fish on its way

---

*For those who like technical terms, this kind of heterochrony is known as *paedomorphosis,* of which Stephen J. Gould and Ken McNamara have distinguished several varieties. The "paedo" root is Greek for childhood. Its mirror image, where new features are added late in ontogeny of more derived species, is known as *peramorphosis.* Again, peramorphosis has been classified into several varieties.

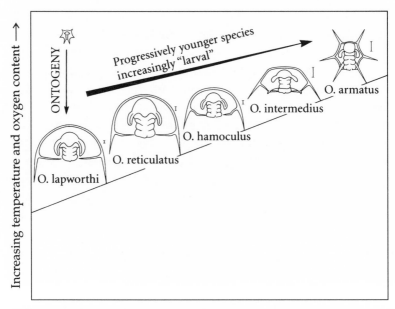

Time shifts: Ken McNamara's diagram showing how successively younger species of the early Cambrian trilobite *Olenellus* from northwest Scotland resemble earlier and earlier growth stages of their ancestral species.

to mammal has long been discarded. The living horseshoe crab, *Limulus*, was supposed to have a "trilobite larva" indicative of its common ancestry with my favourite animals, but this is a resemblance due to a shared simplicity as much as a shared ancestry. But legitimate examples are still to be found in fossil lineages. Among the pelagic, seafaring trilobites I studied, for example, huge-eyed adults actually had *larger* goggle-eyes than their larvae and immature growth stages, which were more like the relatively normal-eyed trilobite from which they had descended. In this way, development timing marched on beyond the ancestor; a good new feature was exaggerated; and what started as a novelty became an institution.

Trilobites can demonstrate seminal facts about evolution. Modern biologists' evolutionary study has disappeared progressively into the genome, and there it has wrought wonderful things; but what is lacking is a time frame, case histories that see evolution in process in real time and real space. Experimental biologists have at most a few years to play with; to a palaeontologist a few million is "but the blinking of an eye." Trilobites can, indeed, provide evolutionary examples worthy of poor Dr. Kaufmann's young life. Changes in development timing, with their profound results in shape, may be the result of no more than tinkering with the genetic code. The molecular finger that resets the clock may do it with no more than an insouciant tweak: even a single gene might control the time switch that creates a difference as profound as those between *Olenellus lapworthi* and *O. armatus*. It is the job of molecular biologists to identify the controlling genes (and I doubt not that they'll still be there tucked away in the DNA even after 500 million years or more), just as it is the duty of the palaeontologist to describe examples of the same genes in action, and how long in geological time and space they take to spin their creative magic.

There is no change without death. I have portrayed the creation of a species, but not its destruction. The history of the trilobites was a history of the passing of the old as much as the appearance of the new. This turnover of species—life after death after life—is the stuff of normal evolutionary change (scientists often refer to it as "background rates"). The better adapted replaced the worse; or the species that originated allopatrically replaced another that shared a common ancestor merely because climate changed to favour an interloper. Life has always been a messy business, and prosperity has alternated between luck and virtue in biology, as in human affairs. Maybe we can turn to trilobites for objective witness of the respective influences of chance and design. Their molecules are lost for ever, naturally. But the signature that their molecules left upon their bodies, which were saved in the geological record, endures till rock crumbles.

Ultimately, trilobites did not cut the evolutionary mustard. They were extinguished without progeny. My hope has faded that, when today's mid-ocean ridges were explored by bathyscape, in some dimly-known abyss there might still dwell a solitary trilobite to bring Palaeozoic virtues into the age of the soundbite. Sadly, there has been no trilobitic coelacanth to astonish biologists, no atavistic survivor who might answer directly all those questions we would like to ask of the genes. Three hundred million years was course enough.

Without death there is little innovation. Extinction—death of a species—is part and parcel of evolutionary change. In the absence of this kind of extinction new developments would not prosper. In our own history, periods when ideas have been perpetuated by dogma, preventing the replacement of old by new ideas, have also been times of stultifying stagnation. The Dark Ages in western society were the most static, least innovative of times. So the fact that trilobites were replaced by batches of successive species through their long history was a testimony to their evolutionary vigour.

Just as mechanisms for generation of new species can be understood in the field and in the laboratory by studying trilobites, so we can map out the reason underlying their slow decline. During their heyday, hundreds of different genera were spread through almost every marine habitat that we know. If you measure success by sheer numbers and variety then the true Age of Trilobites ranged from the middle of the Cambrian to the Ordovician period. But they were fecund throughout their history: even in the latest strata to yield their remains several species can be found together. It is tempting to portray their history as a rapidly building *crescendo* followed by a slow *diminuendo* which lasted until silence finally prevailed. Such an analogy would be misleading. As in much of the story of biological diversity, trilobites prospered and suffered setbacks by turn. Their extinction phases coincided with those that affected many other kinds of animals. These were times when the usual rates of extinction were accelerated, when losers were weeded out and winners favoured to survive and subsequently prosper. Some animals that appeared at about the same time as the trilobites—clams are a good example—endured the seesaws of fate alongside their arthropod contemporaries, and in the end outlasted them all. With trilobites, there were many casualties along the way. Extinction events close to the beginning of the late Cambrian removed many trilobite families that had appeared early in the history of the group. A better studied event at the end of the Ordovician, some 440 Ma, extinguished many more of the families which had given earlier faunas their particular flavour. The tiny, blind agnostids, those enigmatic miniatures from the Cambrian, disappeared. They had lasted nearly 100 million years—reflect on the few millions of years that our own genus has survived, and ponder the meaning of "success." Many large trilobites related to *Isotelus* and *Ogygiocarella* also died out, as did small ones like trinucleids whose medallion shields were so typical of Ordovician strata—in fact, most of those animals Peter Sheldon studied in such detail perished without progeny. Then, too, the free-swimming, giant-eyed, pelagic trilobites of which I had grown so fond are never found again after the Ordovician, and I believe that trilobites failed to occupy that particular ecological niche after that period. Olenids died out, my favourite family from my Spitsbergen days, which had patiently prospered ever since Cambrian times, holding their own against all comers. Truly this was the end of a biological world.

The termination of the Ordovician was also when a great Ice Age, centred upon the South Pole—which was at that time in northern Africa—spread its refrigeration almost completely around the world. Ice ages recur in Earth history, rarely and irregularly, and always with profound effects. The Pleistocene Ice Age with its woolly mammoths and cave bears was merely

The head and tail of *Mucronaspis* (here from the Ordovician of Thailand), a ubiquitous trilobite at the time of the great Ordovician glaciation.

the latest of them. Ice ages yield characteristic rocks, those dumped by retreating glaciers or produced by the fallout from floating icebergs. They share a kind of promiscuous variety: large and small boulders or pebbles lumped and jumbled together, and rocks of different origins all mixed up. Ice simply acts as a carrier, and when it melts everything it picked up along the way just drops. The resulting rocks have a lumpy texture, looking from afar like a badly cooked plum pudding. These characteristic tillites abound in many rock sections containing strata which were deposited close to the end of the Ordovician, and associated with them fossils can very often be collected which are known as the *Hirnantia* fauna. (*Hirnantia* is not a trilobite—it is a brachiopod the shells of which are typical of this Ice Age.) It is astonishing how widespread the *Hirnantia* fauna is. A special trilobite, *Mucronaspis*, is one of its typical denizens, but other trilobites are usually very rare. It is recognizable by a little spike at the end of its tail. I have collected it from a wet and breezy hillside in North Wales, where Cwm Hirnant provided the inspiration for the name of the diagnostic shell. I have collected it again in a humid quarry in southern Thailand, where beads of sweat from my brow blobbed on to the cephalic shields as soon as I wrested them from their enclosing sandstone. I have seen the same trilobite from shales collected from beneath the Tablelands of South Africa. I have seen it from Poland, and Norway, and China. The implication is perhaps rather obvious, but no less interesting for that. These were "cool" trilobites. They chased out others from climates which had been more temperate before the ice sheets advanced—and the effects of the glaciation penetrated to the equator. *Mucronaspis* imposed a uniformity almost as pervasive as the blue suits that blanketed China during the ascendancy of the Maoists. It is now known that extinctions also occurred in the deep seas at more or less the same time as *Mucronaspis* spread over continental shelves, and affected planktonic organisms as profoundly. Many of the trilobites which became extinct probably spent their larval life in the open sea as part of the plankton, and this may have rendered them particularly vulnerable. Through this critical bottleneck, only fortunate trilobites passed. There was no way of knowing in advance that being "cool," or not having planktonic larvae, might equip for survival. These trilobites did not store up the genetic equivalent of ships' biscuits to see them through the hard times. Some simply possessed—by chance—features that would serve them well in the crisis. This is an important discovery about the very nature of mass extinction. Who knows if lessons may yet be drawn from my animals which might influence the actions of another animal—the one e e cummings called Manunkind? And who is now causing another extinction as severe as that endured by the trilobites at the end of the Ordovician . . .

But the end of the Ordovician was by no means the beginning of the end for the trilobites. Families which passed through from the Ordovician abounded in the Silurian—in fact, there may have been almost as many species as earlier, but derived from a more limited set of common ancestors. Crusty-headed encrinurids and spiny-tailed *Cheirurus* grace many collections, and it is tempting to believe that the evolving ecosystem prodded trilobites into yet more inventiveness with their versatile exoskeleton. This was the time when phacopids, with their clever eyes, started to come into their own. Rock surfaces can be covered with them: the Silurian sea-floor could be as crunchy underfoot as it was at any earlier time. Many of these trilobites continued into Devonian strata, which was the acme for all things spiny and blistered, pustulose, scrofulose and carbunculate. But alongside such trilobite extravaganzas were more ordinary citizens like *Proetus* which might, at a glance, be mistaken for an average Cambrian or Ordovician animal. It was *Proetus* and its allies (*Gerastos*, p. 188) that survived the next crises late in the Devonian—the trilobites had by then enjoyed some 80 million years of plenty since the last mass extinction. In some ways, the Devonian events are more puzzling than the Ordovician. There are several of them, one after the other, and each is associated with an invasion of oxygen-poor waters over the continental shelves, which had the effect of removing the coral reefs in which many trilobites lived. This was more like death from a thousand cuts than from a single assassination. The *coup de grâce* was the Frasnian–Famennian event (the name describes its stratigraphical level between two geological divisions), which has been ascribed to a gigantic meteorite impact—the kind of phenomenon usually cited as the cause of dinosaur extinction, a catastrophic event which happened 180 million years after the last known trilobite.

Whatever the cause, after the Frasnian–Famennian only *Proetus* and its allies survived into the Carboniferous. What had been dozens of families had dwindled to a handful, all of them closely related. Even so, many new types of trilobites appeared as innovations during the Carboniferous period. About twice a year I get a parcel of papers from German specialists describing a new batch of species—the discoveries never seem to come to an end. Bob Owens from the National Museum of Wales has found new forms in the familiar crags of the Carboniferous Limestone that make up "the backbone of England," the stone-walled, sheep-studded uplands of the Pennines. Proetide trilobites spread out into many of the ecological niches that had been occupied in earlier times by trilobites from a richer selection of families. They managed to play the same ecological tunes as their forebears, but they used different evolutionary instruments. They spread into deep water, and into newly re-established coral reefs. As a consequence, some of these late trilobites came superficially to resemble their ecological twins recovered from Ordovician, Silurian and Devonian rocks. There were even some species that came to look like *Phacops*—although they did not develop the schizochroal eye . . . What a marvellous dissembler is nature! If I were of a more anthropocentric cast of mind I might wonder if palaeontological puzzles had been placed in the rocks simply to test the mettle of scientific investigators. Biologists and palaeontologists seem to spend so much of their time unravelling the deceptions of nature. Resemblance of shape is everywhere, for ecological necessity dictates form: animals earning a similar living in the wild resemble one another—bat and bird, skink and snake. To pluck out deeper evolutionary truths, the origins of anatomical structures must be recognized—what is termed homology. Homology reveals the deeper concordances of genes and development against the Sirens of overall resemblance. Is this glabella the result of a modification from some deeper design, one which truly reveals common ancestry with another trilobite altogether from one we had supposed at first glance? Is what we see in morphology primarily related to life habits, just as all "flatfish," while doubtless flat, may have come from more than one ancestor? Maybe, while he contemplated *Ogygiocarella*, Lhwyd sensed the important resemblance—ecological equivalence—while making nonsense of true biological affinity. A trilobite may yet be a fish in spirit. In palaeontology, as in human affairs, there is more than one kind of truth.

By the Permian only a modest number of trilobites re-

LEFT: *Gerastos*. Three fine individuals of this neat little proetide trilobite, which almost seem to be saying "two's company, three's a crowd." Large eyes are very close to the glabella, genal spines are shot, and thorax has ten segments. Devonian, Morocco. (Photograph courtesy Prof. Brian Chatterton.)

BELOW: One of the last trilobites, *Ditomopyge*, from the Permian of Wichita, Kansas—two views of an enroled specimen (×3). (Photograph courtesy Bob Owens.)

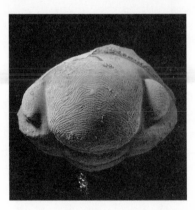

ABOVE: Enroled example of the Ordovician trilobite *Symphysurus* from Sweden, natural size.

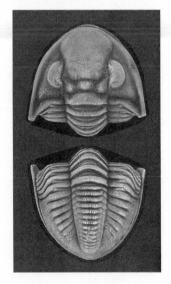

mained, twenty or so genera. Even so, they can occasionally be common fossils. The very last trilobites seem to have disappeared a little before another great mass extinction at the end of the Permian; by then, they were minor players in the marine drama. Their great days had passed. These postscript animals are mostly found in rather shallow habitats in what were tropical seas: perhaps this made them especially vulnerable to climate change. I regret that, unlike some of their contemporaries among the molluscs and brachiopods, none of these late trilobites were adapted to deep-sea life, where they might have seen out the traumas that swept across the land and continental shelves. As it was, they were part of a scene-change that presaged a new act in the story of life. I doubt that we have yet discovered the *very* last species, that rare survivor that still plied its Palaeozoic habits when the ancestors of the dinosaurs were strutting around the streamsides of Gondwana. The trilobites did end with a whimper rather than a bang. I am reminded of the piece that Joseph Haydn wrote as a subtle protest against the mean musician's wages at the court of Esterhazy. In the final movement of the *Farewell Symphony* the musicians leave one by one, while the music continues vigorously to unfold. In the end a solitary fiddle carries on alone—and only then is there silence.

VIII

Possible Worlds

I have spent much of my working life remaking the world. I have pushed half of Europe across half an Atlantic. I have closed ancient seaways and opened up others. I have been able to name an ocean greater than the Mediterranean, and then condemned it to perdition. My job has been to describe the outlines of vanished continents, and to plot the seas around them: in short, to draw a map of the Earth as it was nearly 500 million years ago. To do this, I have used trilobites. When I meet some of my commuting acquaintances on the 6:21 home to Henley-on-Thames they occasionally enquire what I have done that day. I have been known to reply: "I moved Africa 600 kilometres to the south." They usually turn quickly to the soccer page.

One of the first books to open my eyes to the seductions of the scientific method was a collection of essays called *Possible Worlds* by perhaps the greatest of science writers, J. B. S. Haldane. One of the chapters was called *On being one's own rabbit;* and this spirit of experimental adventure was typical. It encouraged me to speculate upon the many mysteries of the world, and how unravelling one or two small ones might be the best thing to do with a life. Now, by a twist of fortune, I am privileged to create my own possible worlds: vanished worlds, written in a geography generated in my imagination, and argued out with a dozen of my colleagues. I have dreamed of chains of volcanic islands belching fumes and spewing lava into archipelagos swarming with trilobites and nautiloids. I have seen these animals suffocate on a ravaged sea-floor, killed and immortalized at one stroke. On a Welsh mountainside I have tested the truth of such an ancient tragedy by breaking a hard rock in which memories of volcanic ash render the surface grey as woodsmoke, and in which lies entombed the shadow of a trilobite, petrified to tell of its dreadful end. In my mind's eye I have seen volcanic archipelagos collapse and die as continent collides with continent, squeezed between masses so vast that an ancient Stromboli might be as vulnerable as a grape in a nutcracker. This is the Ordovician world, a globe so alien that it bears little comparison with the atlas of today. There is land and sea, to be sure, but the continents are not those we have learned by rote in our first classroom. They are strange shapes, curiously arranged.

It is not so long ago, geologically speaking, that our present-day geography was a matter for speculation. In Hereford Cathedral, in the middle of England, the Mappa Mundi is displayed in a dim light for its own protection, but it seems an appropriately mysterious illumination by which to inspect Richard of Holdingham's parchment world of the late thirteenth century. And what a curious construction it presents. With its domination by land rather than sea, the Mappa Mundi looks quite unlike the familiar Mercator projection of our own world. In its centre lies Jerusalem. The British Isles are placed on one edge. But the cathedral town of Lincoln is portrayed with something approaching realism: a street lined with houses runs down to the River Witham from the cathedral on a hill. Like the famous cartoon cover of *The New Yorker* showing a detailed Manhattan Island from which the rest of the world retreats in ever sketchier outline, Lincoln must have been the axis of the known world for the creator of the Mappa Mundi, and the detail beyond was approximate. Travel was difficult; cartography was imprecise (and perhaps Richard was as reluctant to explore as some New Yorkers to venture beyond Brooklyn). At first glance, the lands around the Mediterranean seem impossibly vague, but closer inspection shows Cyprus and Sicily, almost recognizable. In the more remote regions dwell monsters and giants: the satyr in Egypt; people resembling birds (*cicone*) near Samarkand; in India birds called *avalerion* which produce two eggs after sixty years, and then drown themselves as they hatch; unicorns. The accurate cartography of the Renaissance and beyond banished these mythical beasts to ever more remote redoubts. Some would still have them lurking in deep lakes in the Andes, or in remote Amazonia: the last hiding places. In making geographical maps of the Ordovician I, too, am sending dragons packing, firming up vague shapes, and restoring some kind of truth.

The Permian Mappa Mundi, which is the continent of Pangaea, has become almost familiar. The supercontinent that bound all our present continents into one is numbered among those facts that many people tend to pack away in their portmanteau of memorable scientific concepts, like the notion that *pi* can never be exactly evaluated, or that black holes eat matter. The persuasive, complementary shapes of the coasts of eastern South America and western South Africa now seem to make sense: they are a legacy of the divorce of the supercontinent. The South Atlantic ocean widened, progressively, from a fissure to a wide sea, as oceanic crust was added at the mid-Atlantic ridge. Africa and South America moved apart on their respective plates. What once seemed an outrageous idea can now be accepted with a nod—of *course* the continents were once united: it's obvious! India rifted from the eastern side of Africa (leaving Madagascar stranded)—and as it impinged on Asia squeezed into existence the highest range of mountains on earth—the Himalaya. On satellite photographs the wrinkled range seems to crumple before the wedge of the subcontinent, and one can almost feel the pressure that threw up Mount Everest. From the vantage point of space, it seems as if mountains can be made as easily as one may scrunch up a tablecloth by leaning on a place-mat. The Alps, similarly, wind in a line through Europe, a rumpled seam of tectonics that tells another tale of crust buckling under the motor of movement—in this case movement of Africa northwards, shuffling a whole series of plates across the Mediterranean. Pangaea broke up, a marriage *pro tem*, a marriage made not in heaven but in the tectonic basement of the world.

The unity of Pangaea corresponded with the trilobites' demise. Some researchers have sought to relate the former splicing of the continents to major extinctions, and it is unquestionable that the newly annealed supercontinent created unusual conditions to which few organisms could successfully adapt. As we have seen, the trilobites were already vulnerable. But what of the earlier history, when trilobites still ruled the world? (I realise I am being over-emphatic in my imagery here, but just occasionally I lapse from scientific propriety and cock a little snook at the hegemony of the dinosaurs.) For twenty-five years or so it has been recognized that Pangaea itself was but a phase in the history of the continents. Plate tectonics did not begin with the break-up of Pangaea, any more than it has ended with the volcanic eruptions on Monserrat. Rather, the trajectories of continents are the surface expression of the internal engine of the Earth, deep convection driven by the heat of the interior carrying the superficial plates like skin on a cauldron of broth: unstoppable currents, nearly as old as Earth itself. Before Pangaea there were other *Possible Worlds*, other designs for the Mappa Mundi. Pangaea itself accumulated from the collision of still earlier continents: it was but a brief phase of unification preceded, as it was followed, by a longer period when continents and oceans divided the Earth's surface piecemeal. These earlier continental masses came together through tectonic evolution to stitch Pangaea together, like an ill-made quilt. The substance of the earlier continents was the same ancient, Precambrian continental crust that still makes up most of Africa, North America (Laurentia), Siberia or the Baltic Shield. But it was cut into different patches from those we recognize on the school atlas. There was no obligation on the part of Nature to use the same pieces to design an Ordovician continent.

Oceans once separated these earlier continents. The oceans were destroyed little by little as the marriage of Pangaea was consummated. Oceanic crust was obliterated by subduction where plates plunged downwards into ocean trenches; it was the same mechanism in the Palaeozoic as is seen today off the eastern coast of Japan. Ordovician volcanic rocks yielding the remains of trilobites may have been produced around islands comparable to the great volcanoes of Indonesia—these are the explosive expression of plate destruction; the trilobites are testament to a sea troubled by blasts of steam and incandescent clouds of ash.

If the oceans of the Ordovician have vanished, how do we know they were once there? If they had simply disappeared without trace they would indeed be invisible now. But virtually all ancient oceans leave their signature upon the Earth's surface. Continents originally separated by oceans eventually collide with one another and throw up mountain ranges—in just the same way as India's collision with Asia generated the Himalayan ranges. Ancient mountain ranges cross today's continents like old scars. These linear wounds mark the course of the margins of former oceans. Erosion over tens of millions of years has worn away mountain chains of great age, so that they are low compared with the comparatively juvenile Alps or Andes. Look at any topographical map of Asia and you cannot fail to notice the Urals, winding across the vastness of that continent all the way from the island of Novaya Zemlya in the Russian Arctic (where my Oslo sage Olaf Holtedahl established his reputation describing some of the ancient rocks) southwards towards the Caspian Sea. It looks like a seam, and that is exactly what it is: a mountain range marking the seam between a Baltic and a Siberian plate. In the Ordovician, these plates were far apart, an ocean apart: different worlds destined to collide. They became annealed only when the ocean between them had been entirely consumed by subduction—and this unification happened long before the greater marriage of Pangaea. The former existence of an ocean is betrayed by extinct volcanoes of the type associated with subduction, or by volatile minerals and copper ores that leak up from the interior of the Earth when oceans die. Very old plate boundaries may not be so obvious, especially if they have been partly covered by younger rocks. To reconstruct primeval geography the scientist must find and unzip those old scars, open out once more the vanished oceans, running the tape of time backwards, further and further into the past. The more distant the past, the more the uncertainties in positioning any continent, the more like Richard of Holdingham we become. My travelling companions on the Henley-on-Thames train might have asked, with some justice, "Moved Africa 600 km? Why not 900 km? Or two thousand?" For we struggle to know the Ordovician world imperfectly, like trying to solve a jigsaw puzzle through the wrong end of a telescope: a hundred kilometres or so can represent the temporary amnesia of a bad afternoon.

So we have to forget the geography we know, and think afresh of Possible Worlds. There are some tools to help us. Several rock types contain magnetic minerals. The heavy, dark iron ore magnetite was the material used first to investigate the properties of magnetism by William Gilbert, court physician to Queen Elizabeth I, whose *De Magnete* (1600) presciently observed that the Earth "behaves like a giant magnet." The magnetic field streams between the magnetic poles just like the "lines of force" that iron filings trace on paper around a bar magnet. Accordingly, suspended magnets inevitably point to the Earth's poles. Magnetite is a common mineral in nature, often occurring as disseminated grains in sandstones, scattered like seeds in a cake. When a rock is deposited (or a lava erupted), if it contains magnetic minerals they will acquire the magnetization prevalent at the time. This magnetization remains, a fossil of its own kind, even when the plate on which the rock ultimately resides may have moved far from its place of origin. By making some comparatively simple measurements on the angles of inclination and declination of magnetism the position of the pole at the time of magnetization can be recovered—like an accusing finger pointing polewards, the rock magnetism betrays its place of origin. The ancient latitude (or *palaeolatitude*) is revealed by this method, but longitude is much less precisely constrained, so that the position of a given continent is never *exactly* located. But these data provide a wonderful starting point to reconstructing an ancient global geography: so much so that palaeomagneticians are often referred to as "palaeomagicians" by their colleagues, with only a hint of sarcasm. The further you go back in time, though, the more problems there can be, until by the era of the trilobites many of the measurements of the palaeopoles prove unreliable; rocks can be remagnetized later, for example, or the signal can become corrupted. This has led to conflicts between the palaeomagicians and the palaeontologists, each defending their different ancient geographical interpretations. Occasionally, these get to be shouting matches. The palaeomagicians pronounce that only their science is "hard" science, and once I heard one of their number pronounce that "one palaeopole is worth a thousand fossils." I suspect that the same scientist would proclaim that one physicist is worth a dozen palaeontologists—the misguided cad.

The use of fossils in reconstruction of vanished worlds has a long and honourable tradition. Fossils were, after all, key ingredients in the arguments about the reality of Pangaea, and this before many physicists had accepted the idea of a great continent. How could you have such similarity between the floras and faunas of Permian South Africa, South America and India unless they had been once conjoined? Trilobites can be used to rehearse similar arguments: we can use them to map ancient continents. They swarmed in the shallow seas that flooded the interior of Ordovician North America; they abounded where the seas washed over the frigid shores of Gondwana (see page 202); they crawled in soft muds over what we now know as southern Sweden and Estonia. The trilobites despise our political barriers: they follow only the bidding of their own taste in geography. Trilobites that lived in these shallow seas were influenced by climate and environment, just as marine organisms today are different at the tropics and at temperate latitudes. Sea creatures have their own thermometers, and most of them are picky gourmets about what they eat, and where. Predators specialize on prey with the care of a connoisseur sorting out a Château Lafite from a vin ordinaire. Some animals revel in lime; others choose sand as a hiding site; still others dote on sticky, black mud. In short, sea animals have a sense of place, and trilobites were no exception.

When the Ordovician continents were dispersed around the world's oceans the trilobites developed differently on separate plates—especially when they were at different latitudes. Each continent acquired a character of its own—perhaps I should rather say a cast of characters of its own—and many of the characters were trilobites. Map the trilobites and you map the continent. With help from palaeomagnetism it is possible to pinpoint the latitude to which each set of trilobites was adapted. Then, too, different rock types tend to accumulate at different latitudes. If we can recognize an appropriate set of rocks then it is possible to make an informed guess at the ancient environment. Limestones laid down under tropical sunshine are distinctive enough. They often form great thickness of strata which are consolidated from muds of calcium carbonate known as aragonite. Today, you have to go to places like the Bahamas to find their match. Collecting fossils from great cliffs of former tropical limestones can be a dispiriting experience, as your hammer bounces helplessly off the intransigent surfaces. With more experience, you scan the rock face for little, tell-tale signs of trilobitic life—a bit of a pygidium, perhaps, subtly projecting from the rock. You curse the fact that limestone and trilobites are made of the same material, calcite, as you try to lever out a block with your precious specimens somewhere in the middle. I have lost two fingernails this way. But the trilobites in limestone are usually beautifully preserved—if only you can get them out. At the other end of the ancient world, there were no limestones in areas close to the poles. Shales are typical trilobitic rocks, from which whole carapaces can be collected with ease, but they are seldom as beautiful as the limestone examples. Thus sedimentary rocks, fossil species, and palaeomagnetic measurements all contribute to a picture of where any given locality was at the time when the trilobites thrived.

Imagine that you are one of a team of alien geologists visiting this planet 200 million years hence, after mankind's excesses had sterilized the continents as naked as they were in the Ordovician. The engine of plate tectonics would not have stopped, for all our passing. Now imagine that Australia had been split apart—in the same fashion as Pangaea was riven—into three great fragments, which had drifted on their own course to Antarctica, perhaps, Africa and Asia respectively. How could the alien palaeontologist reconstruct the former antipodean continent? She might well start by recognizing the geological integrity of each of the three fragments. Next, fossil collections would soon reveal a strong bond between these dispersed pieces—kangaroos, wombats, possums, koalas, and a whole battery of other marsupials would be recognized as *endemics* shared between the three fragments. Place them close together, and the marsupials had a home to match their family resemblance. Unless the subsequent tectonics had blurred the outlines, it might even be that the three fragments would lock together in a manner as particular as a jigsaw puzzle solved.

So with the trilobites: in this case, *we* are the visitors from the future, and we travel to as strange a world. It may be objected that Australia's marsupials are terrestrial animals, and therefore a better guide to a former continent than animals that could swim across seas. This is undoubtedly true. But the Ordovician was very different from the present day because then the seas extended much further over the continents than they do now. Shallow seas were like evolutionary cooking pots for endemic species. It would be the same today if the sea once more flooded over the vast plains of Australia, seeping into what is now desert and endless scrub. I have collected trilobites in the very centre of Australia, in a spot so remote that even the dingoes were tame, and slunk up for a look at me. In the Ordovician, this same site would have been as remote from the continent edge as it is today—the sea had flooded an extraordinary distance. The dingo looked at me with the same curiosity with which I gazed on a trilobite never before seen by man—we were both aliens in our different ways. From my vantage point on a low hill I could see far away across the peneplain, a place where erosion had done its best, as the book of Isaiah tells us, to make "every hill and mountain low . . . and the rough places plain." It was not difficult to imagine a warm, shallow sea drowning this barren land, and I could reanimate the trilobites in my mind readily enough—a sea thronging with life. In the same rocks we found evidence of what proved to be one of the earliest fish known to science: another alien. Some of the trilobites proved to be as distinctive as kangaroos.

I will now try to draw an Ordovician atlas, my own possible world, a Mappa Mundi of 485 million years ago (see page 201). Some landmasses seem nearly familiar. There is Laurentia—North America and Greenland—united then as now. But it is lying on its side, and the equator passes through its midriff. What is (nowadays) its eastern side is different, too. It has "bitten off" part of the western side of the British Isles. The trilobites from north-western Scotland and western Ireland are the same as those from western Newfoundland and Greenland. The rocks from the island of Skye—to which Bonnie Prince Charlie fled—are the same kind of limestones, precipitated under the gaze of a tropical sun, as are found in New

York State. Conversely, only the *western* part of Newfound-land is part of Laurentia, the Great Northern Peninsula, that long promontory which sticks up like an optimistic thumb on the side of the island adjacent to Canada, where trilobites and rocks tell of connections with Nevada and Oklahoma.

Elkanah Billings, a pioneer palaeontologist in the middle of the nineteenth century, named many of the fossils. His *Bathyurellus* and *Petigurus* were trilobites belonging to a fam-ily, Bathyuridae, which were as typical of the Ordovician tropics of Laurentia as kangaroos are of Australia. Find these animals in the rocks, and you know that the ground on which you stand was part of Laurentia. In Newfoundland, they are found only on the western side of the island; their contempo-raries on the eastern side are utterly different. A suture repre-senting a vanished ocean (called Iapetus) passed between the two sides of the island. In the early Ordovician, the east and west coast of Newfoundland were as widely separated by sea as Brazil and Nigeria are today. The map of the Bathyuridae extends northwards into Scotland and Greenland; Spitsber-gen, my geological cradle, was part of the same Laurentian continent. The telltale trilobites are there in the Canadian Arctic in Ellesmere Island, and in Alaska, and through west-ern Canada, and all down through the western part of the USA into the Great Basin of Utah, Nevada and Idaho—then across through Texas, Oklahoma, and up the western edge of the Appalachians to New York State, where the omnipres-ent Charles Doolittle Walcott first described *Bathyurus*. The labours of dozens of palaeontologists mapped the course of the continent stamped with the unmistakable signature of their trilobites. When I came to work in Nevada, many years after my stay in Newfoundland, I hammered out some of the very same trilobites under the fragrant *piñon* pine as I had first cracked from the hard limestones in the Arctic, while being scolded by a tern whose nest I had approached too closely. This striking similarity proves that in the Ordovician the equator ran lengthwise through North America, rather

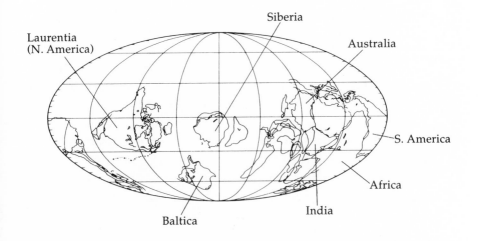

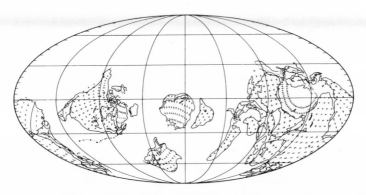

The early Ordovician world, 485 million years ago, as the trilobites reveal, with explanatory map labelling the principal continents. This is a Mercator projection with the ancient equator through the centre. The upper map will help you recognise the present-day continents in their Ordovician positions. The crosses on the lower map indicate the lines of latitude on present-day geography.

than the continent having the north–south orientation of today. (This, I might say, is the simplest of the ancient continents to illustrate.)

At the other climatic extreme lay western Gondwana. The name of the "Land of the Gonds" has a distinguished part in the story of the recognition of Pangaea. The great turn-of-the-century geologist Eduard Suess used it to denote the concordance of geology between South America, peninsular India and Africa (and now, as we know, Antarctica also). They were conjoined in the Permian, and then they were split asunder. But Gondwana existed long before the Permian: it is one of the greatest parts of the Collective Unconscious of the planet. Forged together in the late Precambrian, the basement rocks of Gondwana are more than half as old as the Earth itself: incorruptible, unchanging through half a dozen convulsions which affected vast swaths of the crust. The textbooks that I was brought up with referred to such ancient, stable blocks as "shields" (as in the Canadian Shield) and it is a designation I like because of the connotations of shield as armour, as something that resists attack. In the Ordovician the western edge of Gondwana was close to the South Pole, which probably then lay in northern Africa. The great continent mostly huddled in the southern half of the world, but was so extensive that it stretched all the way from the pole to the equator, which crossed through Australia. No present-day continent is comparable in extent. Another suite of trilobites observed their loyalty to the geography of Gondwana, just as the Bathyuridae had in Laurentia.

A third continent is known as Baltica. On present geography Baltica comprises Norway, Sweden and the Baltic republics—Lithuania, Estonia, Latvia. Eastwards, it extends as far across Russia as the Urals. Recall that this mountain chain marked the former edge of a continent, a seam only annealed when continental Asia was assembled by the collision of Siberia with Baltica. Siberia itself was a separate plate in the Ordovician—all continental seams were unpicked then, all zips unzipped. I explored the Ordovician of Baltica with a Swedish schoolmaster called Torsten Tjernvik in 1975. He guided me around a succession of small limestone quarries in southern Sweden, where the rocks lay horizontal and undeformed—nothing had disturbed them since their deposition about 450 million years before my visit. What was remarkable was how much time was distilled into so little rock. In Wales, I was used to hundreds of feet of dark muds representing a million or two years of sedimentary deposition. In Sweden, half of the entire Ordovician timescale—30 million years or more—could be inspected in a single quarry. A single subdivision of the Ordovician timescale could be as thin as a biscuit: in our jargon, the sequence was *condensed* (deposition was very slow). Yet there were trilobites aplenty, and they were a different set of animals again from those I had collected in Newfoundland. Big tails were everywhere of a creature (related, but distantly, to *Ogygiocarella*) called *Megistaspis*. Not a whisper of a bathyurid. Tjernvik was in his eighties when I visited Sweden. His English was remarkably fluent: he had learned much of his use of the idioms from the novels of P. G. Wodehouse, and the result was charmingly anachronistic. When a particularly fine *Megistaspis* turned up he would say, "absolutely top hole, old bean!" If he wished to impart some important item of information it would be, "can I have a word in your shell-like?"* At the end of the day: "Toodle-pip, old boy!" Everything I saw showed me that Baltica was a separate continent. Both the types of rock and the trilobites (and subsequently palaeomagnetism) suggested that Baltica was located at temperate latitudes, midway between Laurentia and Gondwana in the early Ordovician. As for the trilobites, they were "absolute corkers!"

There is something intimidating about long inventories of names and places, and the capacity to remember such details

---

*Wodehousian abbreviation of "shell-like ear," a poetic cliché originally applied to pretty females.

Ogyginus, a trilobite typical of Ordovician Gondwana. An example from Shropshire, UK. Life size.

and directories can recall the extraordinary but pointless powers of the *idiot savant.* Who really wants to know the day of the week on which 29 February fell in the leap years of the last few centuries? Even lists of trilobite names are tedious. But the patient compilations of lists of fossils from dozens of localities provide the raw ingredients for maps of distribution: these, in their turn, describe the boundaries of former continents. This information could scarcely be of *more* importance. Today lists—tomorrow the world! So, to break my rule about avoiding lists, here is a roll-call of some of the trilobites which are found in the earlier Ordovician rocks of western Gondwana, and only there, and lived in the cool waters close to the Ordovician pole: *Neseuretus, Zeliszkella, Ormathops, Ogyginus, Colpocoryphe, Calymenella, Selenopeltis, Pradoella, Placoparia,*

Merlinia . . . as fine a parade of classical tongue-twisters as you could wish, and I could go on. Each one of these animals is distinctive; taken together they describe half an ecosystem. I mention this list in particular because it saved my scientific bacon.

England and Wales and the eastern part of Newfoundland together comprise Avalonia, a name bearing the flavour of Arthurian romance, but in fact taken from the part of Newfoundland on which St. John's is situated—the Avalon Peninsula. The rocks tell us that eastern Newfoundland and Wales were once a single entity; in contrast to east and west Newfoundland which were separated by the Iapetus ocean in the Ordovician. Avalonia is what is termed a *microcontinent*—a relatively small fragment of continental crust which may have a history of "drifting"—independent of the great continents of Laurentia and Gondwana. Maybe the Arthurian connotation is not so inappropriate after all: Avalonia struck out on its own on a kind of geographical derring-do, and its story is a tale of departures and skirmishes. In the 1980s there was a scientific conflict about the position of Avalonia relative to Gondwana. With my old friend Robin Cocks—a brachiopod expert—I had proposed that in the earlier Ordovician Avalonia was probably part of Gondwana. In support I had a list of Gondwanan trilobites from Wales and Shropshire: *Neseuretus, Calymenella, Ormathops, Colpocoryphe, Ogyginus, Placoparia, Merlinia*. Now the importance of this tally will be evident: carrying such a list how could Avalonia have been anywhere else? And since there was nothing in common with Baltica either—not one trilobite, hardly a brachiopod—we concluded that cool water Avalonia must have been separated from temperate Baltica by another ocean, which, in 1982, we called Tornquist's Sea. (Tornquist was a famous geologist who had worked on the critical area.) This is how I came to name a vanished ocean. Later in the Ordovician, changes in the trilobite faunas told us that Avalonia had rifted off Gondwana and moved northwards over Tornquist's Sea to collide with Baltica. I confess to a slight twinge of megalomanic satisfaction in playing god with a piece of land that is now home to fifty million people.

Conflict arose because a palaeomagnetic "fix" placed Avalonia much nearer the equator and closer to Baltica, several thousand kilometres from our proposed position. As is usual with such scientific rows opinions hardened almost immediately. We were told bluntly that one palaeomagnetic data point was worth a thousand trilobites. We riposted that if Avalonia and Baltica had been so close how come all the fossils were so different—while those of Avalonia were so like those of France, Spain, and North Africa? It became a test case for us; "soft" science versus "hard" science; fossils versus the machines! In the end, the fossils won; *Merlinia* was victorious. Since *Merlinia* was named after King Arthur's magician maybe the fate of Avalon should have been obvious for thoroughly non-scientific reasons. It subsequently proved that the palaeomagnetic "fix" had been flawed, and a later, better one agreed with the trilobites. Today, Tornquist's Sea is marked on all the maps of Ordovician geography. It has crossed over that mysterious line which demarcates what is still theory from accepted fact. Trilobites triumphant. But as Avalonia moved away from Gondwana towards Baltica, Tornquist's was itself subducted away; a new ocean appeared behind Avalonia in its stead. What plate tectonics creates, it also destroys.

But what of Australia, on the eastern side of the vast Gondwanan continent? The western part of Queensland and the adjacent areas of the Northern Territory were also flooded by that pervasive Ordovician sea. When John Shergold and I travelled to this remote area there was only the vaguest notion of what might be in the rocks. This is country of peculiar emptiness. Hardy eucalypts dot a vast semi-desert, where a few beef cattle eke out an existence only if they are watered from wind-driven "bores." The bores often turn dry or run to poison. Paved roads do not exist. Beyond Boulia tracks drive into the nothingness, and there are areas where "gibber plains" of wind-polished stones render the trackways virtually invisible. It is easy to get lost, and I spent much of my time leaning out of the vehicle looking for broken twigs that might indicate the former passage of a Land-Rover in the last field season. The Fierce Snake lives in these wastes, the most poisonous snake in the world, a creature so spectacularly venomous that one of its bites can kill hundreds of laboratory mice. It obviously needs to be an effective predator in this terrain of thin rations—but why so outrageously lethal? After all, snakes do not eat kangaroos. Surely this is the most literal example of "overkill" in nature. The heat is relentless, but there is an exquisite half-hour in twenty-four as the sun squeezes down into the horizon, and the beer can is cracked open and the steak sizzles on the fire, when you would swear that to work here was the greatest privilege a scientist could earn. Those years of poverty as a research student, the poorly salaried assistantship that followed, suddenly seem worthwhile. "I'm being *paid* for this," you say to yourself incredulously. Then it starts to get cold.

Only once did my enthusiasm for the desert suffer a blow. There are very few pubs in the outback and they are sorry and functional places: a plain bar, wooden floors, a flophouse out the back. Station hands collect their pay cheques after months on the job, intending to go to Brisbane for the high life. They often get no further than the first pub. Their money is credited on a "slate," and there they sit—or more likely stand—drinking it all away. After a week or two of this a kind of dopey, surly aggressiveness is the commonest condition: eyes hooded, boredom mired in alcohol. They become what an Australian would call a "ratbag," spoiling for a fight. To go into one of these drinking dens with a "pommy" accent is just the stimulation they have been looking for. "Bloody poms—can't stand 'em," they will announce, fists clenching and unclenching. This is where the Wild West still exists, this remote island within the Island Continent. Fights still settle scores, real or imagined. To a natural coward like myself, this is terrifying. After my first encounter with one of these drunks I spent three hours speaking in a cod mid-European accent to escape their further attention. It was hard for them to develop an attitude to somebody who came from Wallachia.

Australian Ordovician tropical trilobites proved to be different again. Separated by latitude from those of western Gondwana, and by oceans from those of Laurentia, they, too, had evolved their own signature. There were strange animals with lumps all over the headshield that looked superficially like the famous Devonian trilobite *Phacops*—except that closer inspection showed them rather to be related to Dr. Lhwyd's *Ogygiocarella*, and to *Asaphus* (we named it *Norasaphus*). This was a fine example of how trilobites living in similar habitats could come to resemble one another—like different actors donning the same clothes to play identical roles. This phenomenon is known as homoeomorphy. Where we pulled out these trilobites from the soft, limy sandstones, living examples of the same thing were dozing the heat away under the spinifex bushes: marsupial "mice" are mouse-like in design and habits, but they are truly marsupials along with wallabies and koalas. Nature revels in such deceptions. Shergold and I had just split another from early Ordovician rocks of the outback showing the same trick more than four hundred million years earlier.

It would be disingenuous to pretend that trilobites alone reconstructed the Ordovician world, although they were crucial in resolving some of the disputes. Somewhat regretfully, I have to admit that my days of playing with cardboard cutouts of continents are over. Nowadays, information of such complexity has to be handled by computers which can integrate information from many sources, palaeomagnetism, trilobites, sediments and all. Computers can handle all the problems of projection and scaling that are indispensable to make sense of the results: worlds can be swivelled with the flick of a switch. A computer has made the Ordovician Mercator projection in which the Gondwana continent seems so strangely squashed about the bottom of the world (it's the same projection effect that makes Greenland look so triangular on many modern maps). You can understand what Gondwana really looked like if you project from the pole as centre as shown on page 210. To a computer that is routine. But whatever tricks are used it is always difficult to turn a sphere into a plane, and worse still if the continental shapes are strangers to us. Computer reconstructions are only as good as the information with which they are supplied—the adage "rubbish in, rubbish out" applies just as much here as it does to dating agencies. Machines have been known to line up sad mismatches of continents, doomed never to make a successful marriage.

In this chapter I have described the world as it was for a few tens of millions of years during the 300-million-year history of the trilobites. It is almost a snapshot in time—a time slice might be better—but it is a frozen Ordovician moment in a dynamic history of a mutable world, for the continents hardly ceased in their global peregrinations. By the Silurian, 45 million years later, the ocean that had once separated Baltica and Avalonia from Laurentia—Iapetus—had disappeared, subducted away. The great mountain chain—the Caledonides—running through the Appalachians to Scotland, and thence to the mountainous fjord country of Norway, was the consequence of the subsequent continental collision, an engagement almost as dramatic and complex as that which threw up the Alps 250 million years later. Trilobites that had lived apart were brokered into a forced marriage. The faunas changed in harmony with the geography. The eventual reopening of the Atlantic Ocean when Pangaea broke up long afterwards followed—but only approximately—the same seam as had been closed by the Caledonian orogeny in the Devonian. As a result, fragments of the earlier continents were stranded in positions far from their Ordovician homes: now, northern Scotland is on the opposite side of the Atlantic from Laurentia, to which it originally belonged—contrariwise, the two halves of Newfoundland are welded together today when they were originally far apart. Even as Iapetus had

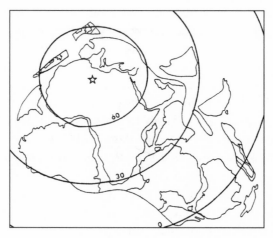

Ordovician Gondwana from a polar projection, centred on Africa, with peninsular India, South America and Antarctica easily recognizable. The southern part of Great Britain is the small promontory at the top of the map.

closed, another seaway—the Hercynian—opened up, running across central Europe and further to the East. We have met this ocean already at the beginning of this book, for it was close to one shore of this seaway that Hardy's trilobite lived (had it not been fiction) and Cornwall's twisted cliffs and noble granites were the legacy of the ultimate demise of that ocean in the next great tectonic cycle. Earth, like a nagging conscience, reopens old wounds. Who knows if some tens of millions of years hence Asia may again cleave apart along the Urals? Who knows if new animals may yet evolve at the bidding of a shattered homeland?

It would take a book as long as this again to relate the whole narrative of the continents as seen through the eyes of the trilobites that swarmed around them. Nearly three hundred million years separates the base of the Cambrian 545 million years ago from the eventual demise of the trilobites. This was a stretch of time that saw the world remade twice. And with each reconstitution of the geography my animals juggled and adjusted to the new climatic and oceanic regimen, sometimes coming together, at others rifted apart. Even now, there are scientific arguments over where this or that great tract of land might have been in the late Ordovician or in the early Silurian. No Mappa Mundi is final, and other worlds are still possible. But there should be enough here to show how geography and evolution have waltzed cheek-to-cheek, and how trilobites give evidence of the changes in partners in the dance.

Now, at last, it is possible to reconstruct the world of the trilobites. We can finally observe the seas that they saw through their crystal eyes. We can understand what Thomas Hardy's desperate hero might have known if a flash of intelligence could have passed between trilobite and man in that mad moment on the Cornish clifftops, a brief vision to strip away the mask of deep time. In the Ordovician, trilobites straddled the globe, from hot tropical seas where corals were already constructing bastions we could recognize as reefs, to cold polar seas where barren landscapes, as yet ungreened, were eroding under the assaults of storms and floods that swept blankets of sediment out to sea to cover the carapaces of our animals, until they at last yielded their secrets to our hammers. We can see vast oceans where none exists today. Across these oceans few trilobites could swim, except for some bug-eyed species which braved tropical storms to spread around the equator, as indifferent to oceanic distances as tuna. Each ancient continent carried its own cargo of trilobites, swarming in their millions. The seas advanced far over these continents, and in the productive shallows specialized trilobites revelled in their place in the ecology; for all its alien setting there were still roles—ecological niches—that would be familiar to us from living seas. (No single trilobite ever ventured into fresh water—if they had, some might still survive.) As it was, there were large trilobites the size of a tureen, like *Isotelus*, which hunted down small "worms" and caused their smaller contemporaries to scuttle away to hide, or roll up into protective balls. Some of these comparative giants grabbed their prey with their strong limb bases and shredded them into pieces, mangling their remains against a fork at the back end of the hypostome. Some species may have been able to stuff their unfortunate prey into an inflated stomach underneath an appropriately inflated glabella (*Crotalocephalus*, fig. 19). Crab-sized *Phacops* may have used its sharp vision accurately to pinpoint its food in dim light. No primitive and unsophisticated mud-grubbers these—they were precision-engineered agents of destruction. There was camouflage, and there was concealment. Spiny trilobites were tight as burrs when rolled up, and just as unappetizing. Others may have decked themselves with small organisms—sea mats or hydroids—the better to conceal themselves in the thronging profusion of the Palaeozoic sea-floor. Others again buried themselves in the soft sediment with only their stalked eyes warily keeping watch by day, to emerge at night to forage among the seaweeds. There were thick-shelled trilobites, which lived close to the tide-line, scurrying in and out at the sea's edge, antennae twitching to the chemical "smells" of food or danger, eyes sensitive to the least movement. These animals would have seen things we shall never see, like tiny animals that have left no fossil to tell us of their existence, or waving algal fronds that decay without trace. Not all history is penetrable.

Wherever the sea-floor was soft, and charged with organic matter, there were true mud-grubbers. Small trilobites, these, like the Cambrian *Elrathia* (p. 243), they searched the sediments for edible particles, incessantly shuffling about the bottom, ploughing through the sediment surface, gleaners and cleaners. They left tracks in a few places, ploughing furrows, braided and scratched by their questing limbs, sometimes flanked by grooves cut by the genal spines. Like footsteps imprinted on a sandy beach, most of the tracks were doomed to erasure, the memory of an afternoon forgotten in the morning tide. But if there were an influx of sand at the right moment they might be preserved—a petrified moment frozen into rock—an occasion when the dance to the music of time left steps behind. Some of these mud-grubbers may have ploughed below the surface layers of the sediment like so many trilobitic moles. A thousand species of shrimp-like animals have the same habits today. These were the foot soldiers of the trilobite world, the *lumpenproletariat*, toiling away incessantly on the sea-floor for a few short seasons. Trilobites with this habit usually have the hypostome mobile, not rigidly attached on the underside of the head, the better to scoop in their squashy and unprepossessing nutriment. They all looked superficially similar, too, all the way from the Cambrian to the Carboniferous, compact little trilobites with genal spines and comparatively small glabellas, and quite a few segments in thorax and tail—they needed limb pairs for sorting out the wheat from the chaff in their diet of slurry. Like the Good Soldier Schweik, they survived when other trilobites, more showy, perhaps, or higher in the marine food chain, failed to survive the extinctions at the end of the Ordovician and late in the Devonian. We have found them with "bites" out of their sides—so some predators evidently found them tasty enough. I might conclude that it is better to grovel hopefully and survive.

Then there were filter feeders. These were generally no bigger than the sediment grubbers, but their headshields were inflated and convex, much more so than the body behind, creating an interior chamber under the head. From the trilobite parade, step forward *Cnemidopyge*, with its frontal spine like a lance in a tourney, and *Trinucleus*, with its strange fringe of doubled-up pits. Sediment was whipped up by the limbs into the head chamber into a fine suspension, from which the edible particles were sorted and ingested. Imagine stirring up a bowl of soup and picking out the noodles. They were sluggish animals, these mud-whisking crawlers, with weak muscles sufficient to propel them from one spot to another to stir up their meagre rations when they ran short. They rested on sled-like genal spines. Many of them were blind, as if their quiet world was not too troubled by predators. But when threatened they could flip their thorax and tail under the vaulted head carapace, tucking away the soft limbs from sight until the danger passed.

Predators, mud-grubbers, and filterers could live together in a single community. Now imagine, if you will, a series of *different* communities of these animals stretching away from the centres of the drowned continents into the deeps that surround them. Progressive depths and different habitats, and in each a host of trilobites did their hunting and scavenging, their digging and searching through sediments, and where mud was soft enough, stirred it into suspension. In deeper environments where the oxygen level was low, specialists like *Triarthrus*, described in Chapter 3, took over from other trilobites in a habitat which diced between plenty and death through suffocation. Above the sea bed, little agnostids swam like animated lentils. At dimmer depths again eyes became useless. This was the territory of the blind, where touch and smell outbid vision, a dark world of palpation and subtle signalling. Around each ancient plate the continental shelves were stacked in order, carrying tier after tier of different trilobites, each minding their own particular business. Now we can begin to understand how there can be so very many trilobite species. Divided by habitat and divided again by geography, trilobites dissected their world into throngs of niches: this is how they became the "beetles of the Palaeozoic."

If we could have sculled over the Ordovician sea it would have tasted as salty, sparkled as brightly under the sun, and been as stirred by storms as the sea is today. On the horizon a smoking volcano might bear witness to the unseen, ineffably slow but inexorable creep of tectonic plates. We would, perhaps, miss the keening cries of gulls, or the silver-sided flash of a shoal of fish. If we threw a deep trawl over the side of the boat, when it is retrieved and tipped out on deck it would have been churning with trilobites. A monster, big as a serving plate, tries to make good its escape by scuttling towards a sluice, its sight bedazzled by the bright light of the surface. Most of the catch would be small beasts—the size of beetles—some of them lying helplessly on their backs, legs threshing ineffectively out of their watery medium. In the bottom of the net there are some balls, round as marbles: a closer look shows that they, too, are trilobites—*Bumastus* perhaps?—tightly enroled against the shock of disturbance. Their protective stance won't do them much good on dry land, but as you lob one back into the water it falls back to the sea-floor, dropping like a stone, and the trilobite finally crawls away unharmed by its experience. Even the mud itself is heaving with tiny trilobites, some as small as ladybirds. These diminutive, blind mud-grovellers are among the smallest of their kind. As you pick through the tangles of weed brought up in the trawl a fantastically spiny trilobite is hidden away in the thicket, an odontopleurid; ouch! you withdraw your probing fingers smartly.

Perhaps we should find out what else is in the catch, besides trilobites . . . Picking through the residue there are some animals that seem quite familiar: a few snails, easily recognizable ramshorns, and a dozen or so small clams. There are shrimp-like creatures, too, and bryozoans (sea mats, colonial animals often forming patches on seaweeds), and a variety of seashells like brachiopods, including one remotely related to some species still living near New Zealand. So not everything is a stranger to us. And delving further into the mud a host of worms of various kinds are revealed—polychaetes and sipunculans—and if we had been able to take a microscope to the mud itself we would have seen single-celled organisms—foraminiferans—and bacteria, which have been processing the waste of the seas since Precambrian times. The Ordovician marine world is a curious mixture of strangeness and familiarity, and we, the fishermen, peek at the nets agog, trying to identify what we know and acknowledge what we don't. The trilobites are lodged in this betwixt and between category, familiar as arthropods, yet strange in all their particularities. If we now row the boat onwards into deeper water, and trawl again, we will bring up another net-load of wonders, another squirming mass of trilobite characters, few of them the same as those from the previous haul. The sea is rich.

The biological world is composed of many small components, yet everything moves together in the great dance of life. The smallest organism has enjoyed a role in the scheme of things, a *locus naturae*. Nature may have been profligate with species, but every species has had a place in the connectedness of things. The small truths of trilobites can be extended to interconnect with a whole world. As a plea for interdependence of culture and science, E. O. Wilson has recently made a case for the unification of knowledge—what he termed "consilience." The trilobite story retailed here shows a consilience of a smaller kind, wherein even identification lists can be married with geomagnetism and plate tectonics to give a portrait of a vanished Earth. The beauty of science is not just the abstract purity of mathematical theorems, which have been celebrated in the biographies of great practitioners like Einstein, John Nash, or Heisenberg, or number theorists and inventors of geometries and algebras. There is no question that reductionist brilliance has yielded some of the greatest triumphs of the scientific intellect. But synthesis can be almost as important as analysis. The attraction of fundamental equations is that they offer hope of an ultimate truth from which everything else may be deduced, even our messy and irredeemably complex world. In following the trilobites we have been looking, instead, at the fruitful marriage of different fields of knowledge, a kind of Pangaea of thought. Or you could think of it as being where different paths converge, like those footways on the Cornish clifftops where Hardy's characters met their pivotal moment, and where the path of the trilobite was joined with the trail of another vanished ocean as revealed in the evidence of twisted shales. My own footsteps, and the account in this chapter, have followed the same pathway. We have explored a past of which trilobites were both witness and victim: and they have been called upon to testify in the reconstruction of their own times and possible worlds. Then by a generous process of reciprocal illumination the world so reconstructed helps us to know more of the trilobite. I see nothing wrong in taking the path to this reconstruction by way of poetic images. In a consilient frame of mind, everything may contribute to an accurate description of the world. I recall two lines from Thom Gunn (*Moly*, 1971):

> *Parrot, moth, shark, wolf, crocodile, ass, flea.*
> *What germs, what jostling mobs there were in me.*

I X

Time

We all struggle with time. Mortality makes time our master, yet we continue to pretend that we can bend time to our will: we *make* time for things, people are said to die *before* their time, as if we all, briefly, had a period when our existence and the time of it coincide perfectly, as with a surfer successfully mounting and moving with the curling crest of a wave. My children ask questions that begin: "In your day . . .?" implying that in some way my time has already passed; was it yesterday, perhaps, and if so why didn't I notice? A palaeontologist has more cause than most to reflect upon time: its measurement, its span, and its consequences. Time can now be measured by the vibrations of atoms to an accuracy which only the leading edge of technology requires. A fraction of a nanosecond is irrelevant to our own lives, and to the pace of a biological lifetime, although it may be germane to the chemical changes that affect a single neuron in the cerebral cortex. Our thoughts are flashes of inspiration, and a flash is brevity itself. However, the duration of a single day is probably our most natural biological temporal unit. When Scarlett O'Hara says, at the end of *Gone with the Wind*, "Tomorrow is another day!" we don't cry cliché, because we all recognize the optimism of a new morning. Witnesses in court are expected to recall a single day; not even an American attorney-at-law would demand a narrative of seconds. The great Argentinian writer J. L. Borges has a short story, "Funes the Memorious," about an unfortunate soul who recalls everything—together with every interlinking ramification—and whose mastery of time has the effect of paralysing him completely. We function thanks to a kind of selective amnesia. This does not release us (especially scientists) from the obligation to speak the truth, a rule which, as we shall see, has been broken even by trilobitologists.

The reader will by now be either insouciant or bewildered in the face of hundreds of millions of years. I have been waving the continents past, ten million years at a shot, with a flick of my wrist. The Cambrian was 545 million years ago; the Devonian lasted for 50 million years. It might be thought that this broad scale pertains to the time of the trilobite; the further back in time the less precision, a few million years is nothing to notice. To the trilobite, Mankind's dominion of the Earth is less than the duration of a single species of their kind. All this is true, yet it is still possible to look into a day in the life of a trilobite—to cheat the great reverse telescope of time which makes distant events seem so small and so far away. Sediment surfaces can preserve a mere day, a true diary of Palaeozoic life. If that day was buried fast enough it may yet be disinterred.

I have already described trilobite enrolment—the instant response to threat which became a time capsule, a moment's panic solidified. Then I have described how trilobites grew by moulting. Their cast-off exoskeletons are testimony to the moment of sloughing off the old coat before growing the new. Sometimes the pieces are cast aside as carelessly as teenage children cast their garments on the bedroom floor. In other cases it is clear that the trilobite adopted a careful strategy for moulting: after all, it was the most vulnerable stage in the animal's life and caution was at a premium. It was not just the hard shell that was moulted: even the finest hairs on the limbs shed their coats at the same time. Where the sea-floor was calm, the cast-off shells are left undisturbed and you may sample the anxiety of the moulting moment for yourself. Imagine, you are sampling a few snatches of time from a larger lifetime, itself a fragment of the time of endurance of a species, which is but a brief instant in the compass of geological time. You can relish the privilege of catching an ancient moment.

Moulting was preceded by a phase in which a special hormone softened the ventral cuticle (see fig. 30); the sutures which crossed the head would then have loosened. When the moment came, many trilobites used their genal spines as levers dug into the sediment to release the free cheeks from the rest of the cephalon (the hypostome was shed at the same time). Since in the majority of trilobites the eye surface was attached to the free cheek this most delicate part was released of its old corneal covering at a helpful early stage. In primitive trilobites this same surface could be shed separately thanks to a suture running all around the eye. The cheeks gone, a gap opened at the front, and the trilobite could then wriggle forwards out of the rest of its exoskeleton, leaving behind a cranidium and thorax to tell of its adventure.

This was often not as easy as I have described, and the thorax will be found separated from the pygidium, or the trilobite will have crawled off with the cranidium still attached obstinately to its head; the trilobitic equivalents of those terrible tank-tops that you can somehow never get out of. There are trilobites that gave the process a hand by inverting their head and scraping off the cranidium; this leaves the cheeks to either side with an inverted cranidium between them—and behind the thorax and pygidium, right way up. In trilobites like *Phacops,* in which the functional sutures on the head have been lost, the whole head often gets inverted, or the trilobites may even have moulted upside down. The observer really feels as if he is watching the most personal gymnastics. Some trilobites may have mated at the "soft shell" stage, as do many living arthropods, and this would have added an extra urgency to the whole procedure. The Palaeozoic sea would have been soaked in hormones—ecdysial, pheromonal, spermatogenic. There are a few examples of "soft-shelled" trilobites preserved, killed before their new carapaces hardened: they have a kind of ghostly quality, a thin and feeble shadow of the real *Phacops*. Some of these animals may have hidden quietly away during this critical stage; my colleague Brian Chatterton tells me of a Devonian burrow in soft sediment (presumably made by some other animal) packed with trilobites in the process of growing their new shells. The burrow served to bury them rather than protect them: the brief time of their tragedy served to ensure the greater time of their survival as fossils.

I mentioned the growth of an individual trilobite when I described evolutionary mechanisms in Chapter 7. Such growth defines a lifetime, the most intimate of all temporal scales, birth to death. It is extraordinary how much we know about the trilobite's personal trajectory. Since trilobites moulted—casting off their stiff exoskeleton and growing a new, larger one—it is an obvious question to trace the same species backwards in time, looking for smaller and smaller carapaces. What was needed was an unusual place where juveniles and adults were incarcerated in the rocks together, undisturbed. Such a locality was discovered "under the pear tree" in one quarry in Bohemia by arguably the greatest name in trilobite research, Joachim Barrande (1799–1883). In what is now the Czech Republic there are remarkably rich sections of Palaeozoic rocks, and Barrande set out to be their biographer. In the Rare Books Room of the Natural History Museum in London a privileged visitor can inspect a shelf-full of large volumes, each bigger than a telephone directory, the fruit of Barrande's own lifetime of labour: the *Système Silurien* * *de la*

---

*Recall that in Barrande's time, what we now know as Cambrian, Ordovician and Silurian were all subsumed under "Silurian."

A trilobite moult, *Paradoxides,* the giant Cambrian trilobite, this example from the Middle Cambrian of eastern Newfoundland. *Paradoxides* is frequently as large as a lobster. Notice the fat glabella and the long spiny thorax with spines extending beyond the small pygidium. In moulting, the free cheeks have been reversed and lie under the rest of the body twisted into this position as the trilobite shed its "old" skin, or exoskeleton, and crawled away forwards. Specimen about 15 cm long. (Photograph courtesy H. B. Whittington.)

Joachim Barrande, the great Bohemian palaeontologist and namer of trilobites.

Bohême. To students of trilobites these are the nearest thing to
holy tracts. Plate after plate of beautiful lithographs (for most
of his life Barrande employed the best artists) delight the eye
today, as they must have astonished his contemporaries (one
is reproduced at fig. 32). It is debatable whether the most
sophisticated modern photography could do better.

   Barrande did more than treat trilobites; he described molluscs and corals and many other fossils besides. However, he
did lavish special attention upon them, beginning in 1852,
drawing upon collections made for the first time from unusually fossiliferous localities. By the end of the nineteenth century every specialist was familiar with the localities we know
today as Šarká and Králuv Dvůr. There is a smart suburb of
Prague known as the Barrandov, where you may have a drink
at the Trilobite Bar. In fact, almost all of this beautiful city is

deeply trilobitic. Barrande himself started his career by accident. He found two pygidia of the trilobite *Odontochile rugosa* close to Zlichov Church during a Sunday stroll. He took them home, but his housekeeper Babinka* threw them out (wives have been known to do this before and since). Joachim made her retrieve them, and his life work was ordained. These two specimens, together with the rest of his huge collections, reside in the Narodný Museum, a grand, columned building which looks down the length of Wenceslas Square in Prague. They are treated with a reverence usually accorded to saint's bones. The visiting scientist will be brought them two at a time, each one immaculately labelled with the original figure of the great man. In his honour a huge plaque was erected on a Devonian hillside in Prague, just one year after his death.

One of Barrande's books describes what we would now call Middle Cambrian trilobites; one plate shows a sequence of a kind related to *Paradoxides davidis*, a trilobite I first encountered in the cliffs at St. David's, Wales, in my early days. In Bohemia, Barrande discovered the whole growth series from babe to adult—a nursery preserved. Since the adult could be as large as a lobster this was a wonder indeed, for the smallest babies were hardly larger than a pinhead. Another species, *Sao hirsuta*, was laid out in even more detail. I visited the famous pear tree near the Bohemian village of Skryje some years ago—it was shrunken now, putting out only a few, sad leaves, I doubt whether Barrande would have recognized it. There were still a few larvae to be discovered in the shales from the quarry beneath it.

Throughout growth, trilobites changed in appearance, but nowhere more so than when they were very small. The trilobite grew—moult by moult—from a tiny individual the size of a pinhead. Barrande recognized that trilobites got larger by the progressive addition of thoracic segments up to the maximum number characteristic of a given species: if a trilobite

---

*Barrande later named a fossil clam *Babinka* in her honour—if having a clam named after you is not, rather, a veiled comment.

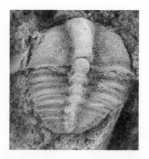

The growth of the Cambrian trilobite *Sao hirsuta* from the Cambrian of Bohemia. The tiny larva, or protaspis, is shown top left. The progressively larger growth stages have additional thoracic segments, until the adult number is reached. The smallest two stages are a millimetre long or less. The specimens illustrated show, progressively, one thoracic segment, three thoracic segments, four thoracic segments, six thoracic segments and thirteen thoracic segments. At the six thoracic segments stage it is just over 2 mm long. The "e" marks the eye position in the larva.

had eight segments, like Lhwyd's *Ogygiocarella*, the babies would add segments one at a time until eight were reached, and after that would continue to get larger moult after moult without any more segments being added. Trilobites were not at all like those scuttling turtles beloved of natural history programmes which hatch out as ready-made replicas of their parents, but changed subtly at each moult. The final number of freely articulated thoracic segments (at which the trilobite is said to have reached the *holaspis* stage) was often acquired when the individual was still only a fraction of its maximum size. The holaspis stage was preceded by a series of moults in which thoracic segments were "released" progressively into the thorax: usually, the smaller the size of the larval trilobite the fewer the segments. Taking this process back to the beginning, at a size of a millimetre or so there were *no* free segments at all: a proto-cephalon was articulated directly against a proto-pygidium with no sign of a free segment interpolated between them. One stage earlier, the first larva of all consisted of a single shield—in which head and tail were combined into a minute disc, termed the *protaspis* (p. 225). In some species the protaspis can be less than a millimetre long. The protaspis hatched out of an egg, no doubt, but claimed fossils of trilobite eggs are controversial. Were it not for the fact that the protaspis is seamlessly connected with intermediate growth stages that connect in turn with the adult trilobite, it is perhaps rather unlikely that these tiny objects would have been recognized as larval trilobites at all. Most of Barrande's protaspides show traces of the trilobite's eponymous "three lobes," particularly the outline of the glabella, even at this minute size (many others don't). But the transformation of a flattish tiny disc into a great predatory *Paradoxides* is a metamorphosis indeed, a life story flushed from the rocks in spite of its immense geological age. It is as intimate a story as the familiar change from caterpillar to butterfly.

It is likely that the earliest growth stages of many trilobites

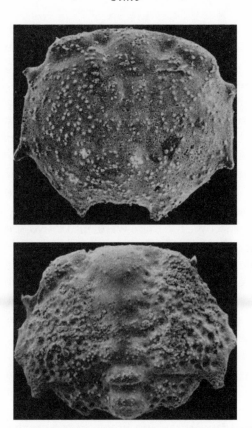

Electron micrographs of protaspis larvae of *Cybelurus* from the Ordovician of Spitsbergen. These single shields show the minutest details, even though they are only a millimetre long. The lower larva is the larger and already shows the proto-head and proto-tail.

were part of the plankton, feeding upon tiny plants, or maybe other larvae, just as baby barnacles or shrimps do today. At some early point in the life cycle the larvae would have settled to assume the beginnings of adult life on the sea-floor. When silicified trilobites were discovered, it was not long before the most beautiful early larvae were recognized among the "fines" in the bottom of the sieve. Fitting the larvae to the correct adults was a matter of skilled detective work, based on finding transitional sequences to a known species. I was lucky enough to find some wonderful protaspides from the Ordovician rocks of Spitsbergen, which are shown in the figure on page 227. These were preserved in calcium phosphate, which replaced the original thin calcareous shells, and so perfect was the replication that tiny spines a few thousandths of a millimetre across are faithfully recorded; just because something is small does not mean it is featureless. Among this cornucopia of microscopic life there were one or two specimens that resembled balloons—but balloons that carried a couple of horns. Harry Whittington had fitted them to an adult that looked very different, *Remopleurides*. Those trilobites related to our old acquaintance *Ogygiocarella* had rather similar smoothed-out larvae—and so, indeed, did *Trinucleus;* no trace on the babies of the fringe that makes the adult so distinctive. My Canadian colleague Brian Chatterton thinks that these tiny lentils were specialized in several ways for life in the plankton. Their undersides were almost entirely sealed in by spiny proto-hypostomes, just leaving holes big enough for three pairs of tiny, thrashing limbs. In freshwater ponds you may occasionally see clouds of minute "water fleas" (cladocerans) beating their way in automatic frenzy through the algal-rich water. My father used to catch them in great numbers and sell them on as fish food in his aquarium shop. My vision of the Ordovician sea and its trilobite plankton is coloured by afternoons spent peering into ponds at a vibrating mist of zooplankton. Unlike water fleas, the flea-like trilobite larvae underwent profound changes and grew to a size that might be a hundred times that of the larva.

Naturally, there would have been no baby trilobites without sex. Sadly, we do not know as much as we would like to know about the sex lives of trilobites. If they were like many living marine arthropods it is likely that the eggs were deposited by females and then fertilized by males. There are several ways of accomplishing this, the simplest of which is for the male to release his sperm into the water where it is free to wash over the laid eggs. It has proved remarkably difficult to recognize the two sexes in trilobites. Nobody has identified any genitalia in the animals that preserve the soft anatomy; nor are there obvious secondary sexual characteristics, like the "claspers" that certain male shrimps use to hang on to females. For most trilobites *la différence* must have been subtle. In 1998 my colleague Nigel Hughes and I recognized what we thought might be the female trilobites of a few species. These shared a peculiar feature: a swelling in the middle of the head in front of the glabella. In some examples the swelling was spectacular. There are living arthropods with such swellings, which are known to function as brood pouches for carrying eggs and larvae. Perhaps this was the trilobitic equivalent of a portable crèche? What gave this explanation an added fillip was the position of the pouch.

A few years earlier I had been choosing my supper in a seaside restaurant in southern Thailand when my attention was caught by the live tank, in which various delicacies are permitted to crawl about prior to being despatched for the table. Among them was a horseshoe crab, *Limulus* or one of its close relatives, slinking dejectedly among the more tasty-looking fish and crustaceans. I was captivated. *Limulus* is the closest living relative of the trilobites (p. 158)—a second cousin, perhaps. Its larva has been known for a century as "the trilobite larva" and does, indeed, have a passing resemblance to the protaspis stage of my own animals. This could be my best chance to find out what trilobites actually tasted like! I ordered the dish. When it arrived I was surprised to find the whole creature had been steamed, and it looked very unappetizing. My amazement increased as the underside of the headshield was lifted outwards—what we would call the doublure

in trilobites—and there *inside* the head was the edible bit of
the animal: big yolky eggs. Horseshoe crabs evidently carried
their eggs in the head region, unlike shrimps and other crustaceans that carry them under the thorax. This was exactly the
same position as the inflated bulbs on the front of the trilobites; circumstantial evidence, of course, but much better than
no evidence at all. And the taste? Even mixed with abundant
noodles it was rancid and intense. I like to think that trilobites
would have tasted sweeter.

The trajectory from protaspis to adult is called the trilobite's *ontogeny*. All complex animals have an ontogeny—our
own, from fertilized ovum through curled embryo to the
developing foetus and baby, being only the most familiar.
Detailed study of trilobite ontogeny has proved unexpected
things. I have already described a mechanism for introducing
evolutionary novelties, by playing around with timing of
development. Many trilobites that are tiny as adults may have
been derived from more normal-sized ancestors by becoming
precociously sexually mature. Study of growth trajectories
revealed this possibility. Identification of early growth stages
showed that some trilobites are more closely related by virtue
of having very similar larvae than would be guessed by a
glance at the comparatively distinctive adults. Larvae can
strip away history to the root of descent. This is not quite the
dictum taught to all zoologists fifty years ago that "*ontogeny
recapitulates phylogeny*"—it might be better expressed as: "by
their babes ye shall know them." So the larvae tell us *Calymene*
is probably related to *Phacops*; *Elrathia* to *Triarthrus*. The trilobite world is still in the middle of unravelling these connections: a new classification of the whole prolific catalogue may
result. It is a wonderful thing to be able to see the *Trinucleus*
fringe develop: it starts as a single row of pits, and gets wider,
and the rows of pits organize themselves into a stippled symmetry. You can watch the spine on *Ampyx* grow through
ontogeny, like Pinocchio's nose under the influence of an

untruth. Harry Whittington discovered that even the earliest meraspis *Ampyx,* with no thoracic segments, could still enroll. As a trilobite, it seems, you could not be too small to need protection. *Odontopleura* and its relatives are spiny even as larvae, a nasty mouthful from the first. No protaspis stage has been discovered for the most primitive trilobites: *Olenellus* and *Agnostus* may even have lacked this stage—unless it was not calcified. They begin their trajectory as earliest meraspides.

As to how the thoracic segments are released into the thorax, I have mentioned that James Stubblefield demonstrated in 1926 that they "bud off" from the front of the pygidium, rather than, say, being released from behind the head. He used the growth series of the small Ordovician trilobite *Shumardia* to show this. *Shumardia* has one extra large (or macropleural) thoracic segment, the fourth one of the six segments present in the adult. *Shumardia* is like other trilobites in the way it grows. It starts as a tiny protaspis shield; then a boundary appears defining the proto-head from the proto-pygidium; then as it continues to moult and grow first one, then two, then three, then four, then five and finally six segments appear in the thorax. During this development the large, macropleural segment appears in the thorax as the last, fourth segment at the four-segment stage of development. At the five-segment stage it is still the fourth segment— and a normal-sized segment has been added on behind it. In the adult, *two* normal segments have appeared behind the macropleural one. In other words, segments are added *behind* the macropleural segment which then shuffles forwards in the thorax as adulthood is approached: segments are budded off the front of the tail. I restudied Stubblefield's specimens sixty-four years after his paper, and found his account exact in almost every particular. Since 1926 his observations have been confirmed on many other trilobites He was nearly a century into his own ontogeny when this book was written in 1999* and was

*He died whilst this book was in press.

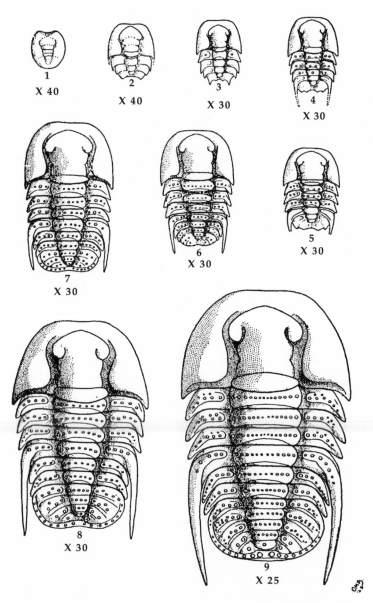

The ontogeny of *Shumardia*. In 1926 Sir James Stubblefield showed how trilobites grew by releasing segments from the front of the tail. The largest *Shumardia* are only a few millimetres in length.

known (even by Lady Stubblefield) as "Stubbie." After his seminal trilobite discovery he rose through ranks of the Geological Survey of Great Britain to become its Director, and eventually was dubbed Sir James. Thus, the trilobitologist who rose the highest did so by making the most minute observations on the smallest of trilobites. In the late 1980s my colleague Bob Owens and I started collecting in the same brooks and dells in Shropshire as had Sir James half a century earlier; from somewhere Sir James rustled up a copy of the original paper he had written on the subject, in 1927. On the cover he wrote, in blue ink: "With somewhat belated greetings . . ."

Then there are trilobite tracks, the imprint of a moment, maybe of one feed in a lifetime. Can fossil time be more fleeting? Assigning a trilobite species to a particular fossil track has proved difficult. Tracks tend to be preserved in the kind of sandy sediments in which body fossils are rare. After all, you would be very surprised to find a dead body at the end of a set of footprints on a sandy beach. Doubtless other arthropods could make tracks somewhat like the trilobite's. So how to finger the culprit? A few years ago in the Sultanate of Oman I was doing fieldwork on some hitherto unexplored Upper Cambrian rocks, perhaps 480 million years old. Trilobites were fragmentary and rare, but since they may have been the only ones of this particular age on the whole Arabian Peninsula they were clearly worth the effort. It was a remote area, the Huqf, almost lacking vegetation except for one, improbable tree. Sandstones and limestones formed low bluffs, so if you followed a single bedding plane you could crawl on your hands and knees over a Cambrian sea-floor which preserved every scratch and footprint. All the signs in the sedimentary rocks pointed to the rocks being laid down in very shallow seas—indeed from time to time the water retreated to the point where it started to evaporate to salt. Few trilobites relished living in such shallow water. The exciting discoveries were beds of beautiful tracks; most unusually, the rocks that

Sir James Stubblefield (right, with pipe) en route to Shetland and Orkney in 1936, with two of his colleagues from the Geological Survey.

filled the tracks actually contained the remains of trilobite shells of one special species that evidently liked inshore life. Could we link the tracks and the animal that made them? If the trilobites made the tracks then they should match for size.

In 1994, I returned to the Sultanate with the great German palaeontologist Dolf Seilacher, doyen of fossil tracks. Together, we spent several days measuring track sizes and collecting trilobites. We gingerly turned over slabs, on the underside of which the casts of the tracks were best preserved. The caution was prudent, because under some of the plate-sized pieces of rock scorpions lurked during the daytime. I learned from my Omani hosts that the big black scorpions, large as king prawns, were less to be feared than the rock skulkers, half the size and yellowish, which carried their stings sidewise, like a lariat. In this empty desert I became aware of the most exquisite connection between these scorpions and the trilobites in the rocks that sheltered them. Scorpions are a kind of arachnoid, that great group of arthropods that includes spiders and mites, together with the most primi-

Trilobite tracks. A cast of the furrows left by ploughing trilobites. *Cruziana semiplicata* from the Upper Cambrian of the Sultanate of Oman. The longest diameter of the specimen is 17 cm.

tive living member of the group, the horseshoe "crab"—
whose eggs I had been lucky enough to sample in Thailand.
Modern studies* of the evolution of arthropods as a whole
have shown that the trilobite's closest relative in the living
fauna is *Limulus,* with scorpions at one further remove. So in

---

*These studies are based on cladistic analyses, as described in Chapter 5.
Such analyses derived from living and fossil arthropods have rejected an earlier
idea that trilobites were closest to crustaceans in favour of their inclusion in a
great arachnoid clade. Trilobites retain antennae, which are lost in the other
arachnoids, or chelicerates. As this book was being completed a new develop-
mental study has shown that the antennae might actually be homologues of the
chelicerae, the curious frontal appendages shared by the living arachnids.

this remote corner of Oman I could pay homage to a meeting across vast stretches of time between ancient relatives: scorpions and trilobites. In the early desert morning, you can see scorpion tracks prodded in ranks across the sand, within a few centimetres of the ancient, fossilized tracks we studied. Dolf assured me that similar scorpion tracks could be found in rocks as old as Devonian. Think of the trilobite *Phacops* still flourishing in hordes at the same time as animals we would readily recognize as scorpions were already wielding their lethal venom, which has so effectively ensured their survival. If they had been able somehow to travel through time and scuttle out of the rock they would have left familiar tracks, alongside those of their living descendants. As I sat on a low bluff in the Huqf glowing in the midday heat I became acutely aware of time past, time present, and time defied.

Our measurements were successful. We found that the size of the trilobites was within the size range of the tracks; and genal spines on the edge of the head of the trilobite were consistent with grooves cut to either side of the digging marks in the furrow. Dolf pointed out to me that this particular trilobite seemed to forage by making tracks with a circular compass—rather like the handsweeps of a window cleaner. The track even had a name, *Cruziana semiplicata*, which had been given to it more than a century earlier by "the English Barrande," John Salter. This trace fossil had originally been found in Upper Cambrian rocks in North Wales, near Merioneth, in wild, wet, mountainous country as different from the Huqf as is possible to imagine. The finishing touch for our theory was that the very trilobite we had fingered as the culprit from Oman has recently been recognized from rocks not far from Salter's original tracks in North Wales. Maybe we are getting as close to a particular moment of trilobitic time as we ever should: one scrape of a limb, one flick of a hair.

. . .

Then there is geological time.

Geological time is calibrated in millions of years, but it can also be calibrated in trilobites. Because they evolved rapidly, trilobites provide a chronometer—a timepiece of personalities. We recognize the faces of trilobites in the same way a skilled numismatist might recognize portraits of minor Roman emperors from a new excavation.

The range through geological time of a species of trilobite is as characteristic as that of King Tutankhamun through historic time. In practice, an investigator hopes to find a whole fossil fauna so that he may cross-check one species with another. I never cease to wonder at how well the stratigraphical system works. My visit to Oman revealed a late Cambrian date for the rocks within a few hours of my arrival. On another pioneering expedition to Thailand I was able to recognize the Ordovician age of some trilobite limestones by comparing their species with those known from China for decades previously—even though the rocks had previously been mapped as Silurian on the geological maps of southern Thailand. These identifications were based on the works of a hundred minor Barrandes—indeed, one of the species from Thailand was first named by one of his Czech contemporaries. Knowledge is cumulative, and hard won. It would be tedious in the extreme to name all the names and pile them in geological order in a great inventory of trilobitic time. There are so many trilobites, and more being added to the list all the time, that a human lifetime is inadequate to know them all. Even after twenty-five years of study there are still parts of the geological column to which I am a stranger. But while geologists and palaeontologists strive for ever greater precision it is still possible to make a grand sweep through geological time to see the ebb and flow of the trilobites, as one replaced another.

In the Lower Cambrian 540 million years ago a variety of trilobites appeared quite rapidly. Those with long, narrow and often tapering glabellas are typical. In North America *Olenellus* and its many allies were abundant, sharing long thoraces with many segments and minute pygidia, often with a very narrow part at the back end behind a particularly large thoracic segment. The long eyes appear to run continuously into the glabella, but there were no moulting sutures on the upper surface of the head. In China, and through much of the Near East, a range of generally similar animals *did* have facial sutures— *Redlichia* is one of the most widespread. I collected fragments of this genus on a relentlessly hot rocky slope in Queensland, Australia, reflecting at the time that this would be the nearest I would ever get to roast trilobite. The Chinese divide their rocks into fine time slices using different forms of these animals as indices. In rocks of similar age appear the first miniature trilobites, such as *Pagetia,* which I regard as primitive relatives of *Agnostus.* Some of these diminutive forms—usually about the size of a small woodlouse—have inconspicuous eyes, and three thoracic segments; others are blind and with two, like agnostids. They carry on into the Middle Cambrian, by which time true, blind agnostids are often abundant. All these small trilobites are outstandingly useful for dating rocks, since some species are very widespread and they evolved quickly into other species. If they were indeed planktonic, this might account for their ubiquity. They are tiny timepieces, and as precisely engineered, for they enroll into perfect little capsules, and every bit of the skeleton collaborates in their protective stance.

Alongside a variety of agnostids Middle Cambrian rocks yield giants like *Paradoxides,* which certainly stalked the sea-floor rather than cruising far above it. Trilobites of this type also provide useful chronometers: if you can recognize *Paradoxides,* you can recognize the Middle Cambrian. There were trilobites of usual size, too, some of them blind, like *Meneviella.* Several of these blind trilobites were known already to Joachim Barrande more than a century ago, and their discovery from rocks in France, Wales and Spain provided an early proof of the affinity of time and trilobite; they could evidently be used to date rocks! Their relatives had normal eyes, so it is likely that these trilobites became

Paralbertella bosworthi, 6 cm long, a typical Middle Cambrian trilobite from British Columbia, Canada.

blind, rather than being blind from their evolutionary birth. They lived in deep, or at least turbid seas. Their oculated kin include *Ptychoparia striata*, a species known from wonderful material in Bohemia. It is, you might say, a Joe Average kind of trilobite, with a moderately small pygidium and a somewhat tapering glabella with medium-sized eyes, and fairly numerous thoracic segments: nothing exaggerated in any direction. The commonest of "rock shop" Cambrian trilobites, *Elrathia kingi*, is a North American equivalent, smaller and wider, perhaps, but just as middle-of-the-road. There are many dozens of broadly similar trilobites, the naming of which causes even the most patient of specialists to gnash their teeth. Their correct determination requires skill and experience, for many similar-looking beasts range through Middle and Upper Cambrian strata. It is much easier to recognize some of the spiny trilobites which appear among the Middle Cambrian faunas, the first of many imitations of pincushions in trilobite history. Trilobites with rather large pygidia became common at the same time. Most conspicuous among these are *Corynexochus* and its allies (see *Paralbertella*, p. 239), a distinctive set of trilobites with long, forwardly-expanding glabellas shaped like pestles, often with spiny thoraces and distinctive hypostomes. The Middle Cambrian was a rich time for trilobites, and when you remember that there were many additional kinds of arthropods in the Burgess Shale, it might be appropriate to dub the Cambrian the "Age of Jointed Legs": it was probably the time when scrabbling limbs were at the acme of design.

Agnostids and many others carried on into the Upper Cambrian. Peculiar trilobites, more or less related to *Damesella* (see *Drepanura*, p. 243), are found in strata of this age in China: they all have tails with different arrangements of marginal spines, looking like combs or strange agricultural instruments. One of them—featured in Chinese medicine as the "swallow stone"—was ground down and incorporated into potions. Most Chinese remedies I have come across are said to be good for old age, and it seems conceivable that this most ancient of medicines was a case of sympathetic magic. Their appearance in the pharmacopoeia made *Drepanura* one of the earliest Chinese trilobites known in the West. Related species extend into Australia, where the distinguished Estonian *émigré* Alexander Armin Öpik described a whole range of special trilobites retrieved from among the prickly spinifex bushes in central Queensland. His North American equivalent is Allison R. Palmer, known to all trilobitologists as "Pete," one of those people to whom the overworked epithet "indefatigable" truly applies. His works on the Great Basin—that vast area of Basin and Range embracing much of Utah and Nevada—are a tribute to mind and hammer over matter. I have climbed some of the same slopes, at an elevation which leaves you gasping. If you are not careful you can slide off on a scree slope all the way back down the way you came. The air is fragrant with conifer resin. *Opuntia* cacti snag you unawares and there is the occasional startling buzz of a rattlesnake, but mostly this is benign, if hot, country, and from Pete's slopes you can see sagebrush stands in the basins dotted with a few cows, and at the lowest point maybe the glistening white of a salt pan. Pete collected all the rocks that crop out on the range-sides, where the trilobites provided another narrative of late Cambrian time, hundreds of different species teased out from the flat-bedded limestones and shales, telling a story of evolutionary radiations and local extinctions. Pete knows every particular of these animals with the kind of relentless enthusiasm that is a distinguishing characteristic of many American intellectuals, and no doubt one of the reasons why they have so conquered the world.

Around the edge of the North American continent, and in Scandinavia, as well as in Wales, the late Cambrian strata yield up different trilobites, Olenidae, which I have already celebrated in Chapter 7. They preferred a specialized habitat, low in oxygen. Geological episodes as short as half a million years can be recognized using these trilobites as timepieces. This may not seem like precision, compared with 4:39 a.m. on 15 August 1931, but 500 million years ago it is precision to one part in a thousand. In time, precision is relative.

Ordovician time was probably when trilobites lived in the most places and occupied the most varied ecological niches in the sea. They lived everywhere from the shallowest sands to the deepest-water shales; in sunlit reefs and in gloomy abysses. Some Cambrian families survived, like Olenidae and agnostids, but what gives the Ordovician its special flavour is the appearance of a whole range of trilobites that formed the foundations of the subsequent history of the whole group: cheirurids, odontopleurids, proetids, calymenids, encrinurids, lichids, phacopids, dalmanitids—the list goes on. I might apologize for giving such a litany of families (typical generic examples of most of them have already been mentioned in this book) were it not also an almanac of Ordovician and later times. A familiarity with a few dozen of these animals will help you peg down time, give you a chronology for the splitting of continents or the appearance of the first scorpion. The names themselves really do matter. Most diagnostic of Ordovician strata are trilobites that did not survive into the Silurian. These include the free-swimming trilobites like *Cyclopyge* and *Carolinites*; while on the sea-floor thrived many asaphid relatives of Lhywd's *Ogygiocarella*, together with exquisite trinucleids, and lance-bearing relatives of *Ampyx*. There were trilobites as spiky as porcupines and as smooth as boiled eggs; trilobites larger than lobsters and smaller than gnats. Because the continents were dispersed at the time there were different trilobites on separate continents—and each one with a chronological narrative of its own. The reader might begin to understand the sense of awe that a researcher feels at the magnitude of her task as she makes a bid to know them all.

At the end of the Ordovician there was a major, or mass, extinction, one of the most important such events in the whole history of life. A great glaciation centred on North Africa undoubtedly drastically cooled the climate in later Ordovician times, and this was probably the main cause of the faunal crisis. You can find the kind of deposits associated with ice ages in Africa and elsewhere—and, remarkably, trilobites quite close by. A few species were evidently tolerant of the cold, and one of these, *Mucronaspis*, spread very widely in the cool era. It was one of the trilobites I collected in Thailand, and it was with no little astonishment that I identified it there with a species originally described from Scandinavia. Trilobite time really works across the world! There was a big loss of families at the end of the Ordovician, and several of those that died out—like agnostids—had a history going back to the Cambrian. Some of my favourite trilobites were among the casualties: no more *Trinucleus*, no more *Isotelus*. I doubt whether any

Elrathia, a typical mid-Cambrian trilobite a few centimetres along at most, with a "flower-pot" glabella, thirteen thoracic segments and a moderately small pygidium. There are hundreds of basically similar trilobites. This specimen is from Utah.

The tail of *Drepanura*, the "swallow stone" from the Cambrian limestones of Shandong, China.

free-swimming species ever reappeared. The trilobitic world after the Ordovician was a different one. But the survivors among the trilobites soon bounced back, and by the middle of the Silurian these remaining families had diversified mightily. A little practice allows the student to recognize Silurian *Balizoma*, *Calymene*, *Proetus* or *Ktenoura*. They are still common enough to function as useful chronometers.

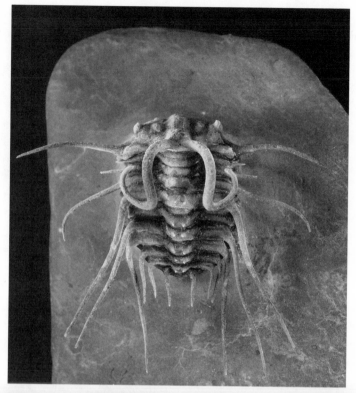

Dicranurus, a spiny relative of *Odontopleura* from the Devonian of Morocco. Life size.

There is much less distinction between early Devonian trilobites and those of the Silurian than there was between those of the Cambrian and Ordovician, or between Ordovician and Silurian. The Devonian was the apotheosis of *Phacops* and its relatives: for a while the schizochroal eyes had command. There was also a remarkable variety of spiny trilobites. In some localities, particularly in present-day Morocco, it seems that almost every Devonian trilobite became covered in prickles and spikes. Just a week before writing these words I saw an unnamed trilobite which carried a great trident sprouting from its glabella, an adaptation as unique as it is inexplicable (see p. 262). Find it again and you would know *exactly* with which moment of geological time you were concerned. Otherwise, the trilobite was not so exceptional, just another relative of *Dalmanites*. Then there are species with great, curled ramshorns originating at the neck as shown in the figure on page 244, or intimidating batteries of vertical spikes. Some trilobites related to *Lichas* decorated themselves with the splendour of a medieval pope; other odontopleurids—always spinulose from their Ordovician origins—transformed into little more than bunches of needles. Even the zoologist hardened by decades of exposure to Nature's curiosities whistles through his teeth when he sees these animals for the first time. Such armour was, no question, protective. Was there some new threat which prompted such an exuberance of spinosity? Could it be connected with the rise of jawed fish at about the same time? As with all such apparent correlations, it is very hard to know whether a particular observation unequivocally leads to the truth. There is usually some alternative explanation waiting in the wings. For an appraisal of time, we do not even need to know *why:* we can treat the bizarre trilobites as one might curious statues or totems from a lost culture, they typify particular moments, they fix the past.

Nearly all these fantastical trilobites failed to survive the Devonian, for later in this period there were a series of extinction events which picked off one after another of the trilobite families. The greatest of these events was the latest, the Frasnian–Famennian event. Very few trilobites survived: even the phacopids were doomed. Those that did survive were all related to one of the less spectacular trilobite groups of the Devonian fauna: *Proetus* and its relatives. Small and compact, they mostly eschewed the spinose extravagances of their contemporaries. Some of them had knobbly heads, but that was about as adventurous as they became before the Carboniferous. By then, time was written wholly in variations of proetoids. My friend Bob Owens will extol their varied subtleties; Gerhard and Renate Hahn are German professors who know every nuance of these trilobites. And it is true that in the earlier Carboniferous, while tropical seas flooded much of Europe, the proetoids produced many different designs. Some of them resembled trilobites from older periods, probably because they adopted similar life habits. There were blind inhabitants of deep waters; in crystalline limestones we find animals that an unwary observer might mistake for *Phacops;* there were even some that look like Ordovician *Harpes.* However, solid-looking, compact, big-eyed, but still smallish trilobites like *Griffithides* were probably the commonest type. My belief is that, far from being in decline, trilobites kept their capacity for invention, reinvading old habitats, and spreading back into deep water. New forms evolved so fast that they are still useful in characterizing chunks of geological time, although it would take an optimist to claim that they were as common in Carboniferous rock outcrops as they were in Silurian ones. Hardy's hero would have been more likely to have had a trilobitic encounter had he dangled from an Ordovician precipice. Clearly, their realm was shrinking; and more so in the succeeding Permian. A few famous localities in Sicily and Timor show that in places we might still have seen a heaving mass of trilobitic bodies had we waded through the shallows there 250 million years ago. New genera still appeared, so that even close to their end trilobites were still capable of ticking off geological time. But then, after calibrating time for a period three times as long as the dinosaurs, the trilobite chronometer stopped.

I should not give the impression that this great temporal story was easily read, as if one massive and continuous pile of strata had yielded one trilobite after another, plucked in order. There are few areas that can be read that simply; it was more a question of pasting together the timescale from snatches of narrative, here and there. There were mistakes and there were arguments, some of them virulent. Even the great Charles Walcott erred. In 1883 he wrote, in his dry fashion: "Below the Potsdam Sandstone [a formation in Nevada] there occurs a distinct fauna, characterized by a considerable development of the trilobite genus *Olenellus,* a genus that in the embryonic development of several of its species proves that it is derived from the *Paradoxides* family and is consequently of later date." *Paradoxides* you will remember as a Middle Cambrian guide, while *Olenellus* was a Lower Cambrian index; he had them the wrong way round. Walcott had evidently confused two of the categories of time I have considered in this chapter: development time of the individual, and geological time. He had plausibly observed that smaller *Olenellus* look rather like *Paradoxides,* but what we now know of heterochrony—time shifts in development—allows us a different explanation. The similarities are only the result of an ultimate common ancestry. Beware assumptions about time, for the rocks will put you right. And this is what happened to Walcott a few years later, when the Scandinavian geologist W. C. Brøgger proved that rocks carrying *Olenellus*-like animals in Norway were clearly overlain by *Paradoxides.* He had found them preserved as consecutive narrative pages, in an undisturbed rock sequence. Walcott, like his friend G. F. Matthew—who worked in folded rocks in New Brunswick where the evidence was ambiguous—had to re-examine the evidence and, like the good scientist he was, he allowed the facts to change his view. He did not persist in attempts to twist time to fit his own preconceptions.

Arguments about the timescale will not stop; as knowledge grows, so the arguments become focused on ever smaller intervals. I have spent much of my scientific life observing slow progress towards an international agreement on how to define the boundary between the Cambrian and Ordovician, a process in which trilobite timepieces have played a part. Esoteric though this problem may seem, I have seen grown men glow incandescent with rage over this metaphorical millisecond in life's history. Candidates for the place to define this time boundary at a particular point in a rock section have been proposed at various locations in Newfoundland, and in Utah, in China and in Norway—I have visited them all.

In China, near a small town called Changshan, I came face

to face with another manifestation of time. Our party was investigating rock sections which spanned the contentious boundary, collecting trilobites from the critical interval while sitting on a warm hillside, happy as a lark (probably happier). From time to time, huge buzzing things flew past—I was rapt enough hardly to notice them. Suddenly, I felt a searing pain in my side. One of the giant insects had crawled up inside my field jacket, where it had doubtless been chafed and irritated by my vigorous hammering. I leaped to my feet, and the largest hornet I have ever seen in my life fell to the earth. It was a mystery how a creature so enormous and filled with venom could manage to get off the ground, let alone fly around the place. As the pain intensified I attempted to interest our interpreter in the urgency of my case. As she didn't know the word "hornet" I flapped my hands and buzzed, doing a dramatic stinging motion into my side. "Ah," she said, smiling warmly, "bee! Not very derangerous!" By now the paddy fields were swimming before my eyes. Luckily my friend David Bruton had seen the whole thing, and eventually made it clear to our hosts that it was one of the flying monsters that had pierced my abdomen. A very strong American called Jim Miller then carried me delicately along the narrow earth dykes that separated the paddies, winding in a kind of rectangular confusion hither and thither. To escape to the road is an intelligence test. In a moment of lucidity I recall staring from Jim's back into a small pond full of water-chestnut plants, and thinking: "my time has come." This, my own personal time, had apparently run out in the middle of China in pursuit of a problem of interest to me and a few dozen others. This was my connection with the search for another moment 489 million years earlier. For a moment, I understood my insignificance in the face of geological time.

Fortunately, I made it to the field vehicle, which carried me rapidly to Changshan. The appearance of a westerner in this remote town in the early 1980s was something of a sensation, and the whole place turned out to follow my prostrate body to the "hospital"—which turned out to be a simple building with no glazing in the windows. Several dozen heads craned through the window for a good view. They were having the time of their lives. I remember little, but I am told that the now tumid swelling was cut with a sterilized knife and squeezed enthusiastically, resulting in satisfactory gouts of blood. A poultice of mashed-up vegetables was then applied. Nodding confidently, my doctor said, through the interpreter: "Here, we use a combination of traditional Chinese and Western medicine." I was given an aspirin (that was the Western part), and a vast phial of big pills made from weeds (the Chinese part). They worked, for in two days I was up and about. Curiously, the incident caused me a certain loss of face. The famous old Chinese Professor Lu Yen-hao said to me when I had recovered: "I have seen these insects many times, but you are the first person bitten." Then, having thought for a moment, he added: "except maybe some peasants." When I returned to London I told the hornet specialist in the Natural History Museum of my experience. "Wish you'd brought back the bloody hornet," he said. "I don't think we've got one of those in our collections."

Science depends on honest reporting. It would not matter greatly if any of the items in the previous story were exaggerated, or even if part of the account was made up for the amusement of my readers—although I should reassure you that I recall everything as clearly as I can. But what no scientist is allowed is deliberately to mislead. It is worse if this deception is in the service of self-aggrandisement. This was apparently the case in the *"affaire Deprat."*

Jacques Deprat was a young geologist employed by the Service Géologique de l'Indochine in the early years of the twentieth century, at the time when what is now Vietnam was a French colony. This was an heroic period for geological exploration. The scientific method had unravelled many of the complexities of the Alps and the Himalaya; indeed, it seemed as if the whole structure of the Earth might be within the grasp of a mind sufficiently bold. Exploration of hitherto unknown territories was a central part of this quest. Jacques Deprat was, without question, a talented and courageous geologist of great energy. He was an accomplished alpinist who revelled in collecting information from inaccessible peaks; a synthesizer with a gift for reconstruction of complex rock structures in three dimensions; and a palaeontologist of some ability to boot. Generalists of this kind are virtually unknown today. He was also somebody who had pulled himself up from modestly bourgeois beginnings by virtue of talent and hard work: a hero for our times, one might say. This was no small achievement in a France that was class-ridden and elitist. To make his reputation, Deprat had been compelled to find work at the edge of the French empire, and even there those who occupied establishment positions—all, like his nemesis and superior officer Honoré Lantenois, products of the French elite education system—despised outsiders. But by 1912 Deprat had a worldwide reputation; his relentless hard work in the field had shown that structures recognized in Europe could be applied to folded and thrust rocks in Indochina, and, most remarkably, that rocks of Ordovician age could be reliably dated from this remote area by the trilobites they contained. They included species named for the first time by Joachim Barrande from the strata around Prague: and what better evidence than species described by the greatest time-keeper of them all? These species are known today as *Deanaspis goldfussi, Dalmanitina socialis* and *Dionide formosa*. The first two named are familiar fossils from the late Ordovician Letna Formation in Bohemia, where they are particularly common trilobites; most old collections have examples tucked away in the back of a drawer. *Dionide formosa* is a somewhat rarer form from the Vinice Formation, but well-known nonetheless. It was evident that these reliable, old-established trilobite chronometers were capable of wide distribution. The formal descriptions of the specimens from Vietnam were made by Deprat's colleague, the in-house palaeontologist Henri Mansuy, in papers published by the Service Géologique in 1912 and 1913. Deprat's reputation seemed to be unassailable.

But then the doubts began. Mansuy began to treat Deprat with caution. Lantenois went further: he asserted that the fossils of *goldfussi* and *socialis* were indeed what they were supposed to be, but entirely because they were, in fact, Bohemian specimens falsely claimed as coming from the locality of Nui-Nga-Ma, near Vinh, in Indochina. They were "plants," falsehoods, cheats; at the very least, to use the delicate word employed by Jean-Louis Henry in his 1994 account of *l'affaire*, they were "apocryphal"—of doubtful authenticity. If correctly identified, such duplicity broke the golden rule of honest science. Deprat defended himself vigorously, and pooh-poohed the charges as slander. He may have accurately intuited that the less able, if better connected, Honoré Lantenois resented the rise of this upstart in the geological firmament. After all, it was he, Lantenois, that had placed the Service Géologique de l'Indochine on a firm scientific footing, but Deprat had had much of the glory. Never underestimate the power of bile. But as one official investigation gave way to another tribunal, as the Société Géologique de France became involved, and the great voices of the day were called upon for their opinions, things looked bleaker and bleaker for Deprat. An official expedition to Nui-Nga-Ma which had been convened to duplicate his findings yielded nothing definitive. Deprat refused to make his field notebooks available to the Commission of Inquiry, behaviour which was decidedly incriminating. Even while tens of thousands of young Frenchmen were being slaughtered in the trenches French justice ground slowly onwards—for the colonial world of Indochina was in another realm of time. Letters took many weeks to arrive by sea, to deliver the latest instalment of justice from the mother country. Time was accelerated for the dying, while Deprat's pursuit by Lantenois was conducted in slow motion, at a distance.

Finally, Jacques Deprat was disgraced, demolished by the same geologists who had once built him up. Professor Termier, his one-time champion and doyen of Parisian geological society, was the reluctant executioner of his reputation. A grand commission of the celebrated Société Géologique de France concurred that the trilobites from Nui-Ng-Ma were apocryphal. Those who had once praised the young Deprat were now the buriers of his reputation. In November 1920 he was dismissed. You must not lie about trilobites, nor yet about time. Falsehoods will find you out.

Except that the obliteration of his reputation did not happen—quite. Jacques Deprat wrote an account of the whole business, after a period in retreat licking his wounds. He wrote it as a *roman à clef* under the title *Les Chiens aboient* (The baying hounds), and there you might find—little disguised—the story of his arrival in Vietnam, of his relationships with Lantenois and Mansuy, and a recitation of his downfall. Of course, it is a partial account, but it does have some ring of truth. It is impossible for the modern reader not to sympathize with the outsider, tarred by an intolerant and privileged establishment. In 1990, M. Durand-Delga made a bold attempt to rehabilitate Deprat at a special session of the Société Géologique de France, apparently accepting the notion that he was "set up" by those jealous of his reputation. This was the eventual defence that Deprat himself adopted (after changing his mind several times). It makes a good, contemporary, psychological thriller. Nor is there any doubt about Deprat's real achievements in other spheres of geological science. And what other scientist, so vilified, might have the talent to play novelist?

But the question remains: did he or didn't he? It scarcely seems possible that Henri Mansuy, elsewhere an exemplary palaeontologist, would in this case relax his standards of probity out of pique. Besides, it was Deprat himself who photographed the contentious cranidium of *Deanaspis goldfussi*, published in 1913. And if he was innocent why did he collude with the suspicions of his enemies by refusing access to his field notebooks? Equally, one might wonder why, since his career was in the ascendant, he should endanger it by so foolish a deception. Was he revelling in a feeling of omniscience? Did he feel so inecure that he must embroider the truth to make it more appealing to the world? Whatever the answer, Deprat was ruined as a geologist.

Several years later Deprat resurfaced in another life under the new name of Herbert Wild, novelist. It was Herbert Wild who wrote *Les Chiens aboient*, and an intelligent observer might have readily made the connection between Wild and Deprat. He enjoyed critical success with several subsequent novels, one of which was nominated for the Prix Goncourt, and earned a sufficient living from his pen to support his family. He also returned to his first love, the mountains, and became a considerable alpinist, a master of the Pyrenees, and pioneer of several of the more challenging peaks. So far as we know, he never wrote about geology again. The mountains eventually claimed his life in March 1935; curiously, he had written a novel which described in some detail the circumstances of the fall in which he subsequently died. Only upon his death was the connection between Wild and Deprat finally revealed.

There is an intriguing symmetry between this story and the one with which this book began—the episode in *A Pair of Blue Eyes*. Both concern fictional trilobites. Hardy's dangling hero faced death reflected in the eyes of a trilobite; a fall killed Deprat, after trilobites had ruined him. In Hardy's case a plausible fiction, a Carboniferous Cornish trilobite, was used for dramatic purposes by a novelist whose reputation is probably as high now as it has ever been. No one would claim that his fame is undeserved merely because the trilobite was a creature of his imagination: it is the novelist's job to whip up such fancies. Deprat was disgraced because his fictions were created under a different convention; he was supposed to follow the rules of the scientific method. However we may empathize with the tragedy of brilliance wasted, or recoil at the vindictiveness of his persecutor, Lantenois, we know that the whole scientific endeavour depends on not doing what Deprat is alleged to have done. There is no possible compromise on this principle: no scientist can be trusted who tells the truth 78 per cent of the time. How do we recognize the flawed percentage? By an exquisite twist, Deprat became a novelist, and one who may even have admired the works of Thomas Hardy. If, as Mr. Wild, he had called upon trilobites to play a fictitious role it would have passed unremarked. The difference between the creative roles of science and art could scarcely be better delineated than in this Tale of Two Trilobites. The distinction is this: like all artists, Hardy made his own time—the compass of the novel has its own bounds which the reader volunteers to enter when she engages with the book. The veracity of the trilobite is incidental, just as it matters not a whit that Hardy thought of the creature as a stony crustacean. By contrast, Deprat's avowal of time was an oath taken on the credo first laid out explicitly by Francis Bacon in *Novum Organum* (1620):

> But if any human being earnestly desire to push on to new discoveries instead of just retaining and using the old; to win victories over Nature as a worker rather than over hostile critics as a disputant; to attain, in fact, clear and demonstrative knowledge instead of attractive and probable theory; we invite him as a true son of Science to join our ranks.

He who abuses the call to "clear and demonstrative knowledge" by the employment of deception is no "true son of Science." Imagination may provide the source from which both artistic and scientific genius flow, but the artist delights in fabrication, just as the scientist revels in discovery. Time tests the quality of the artist's vision, just as it tests the durability of scientific revelation.

X

Eyes to See

Most scientists work in small arenas. To look at popular accounts of scientific discovery you would think that every man or woman in a white coat was seeking to solve the problems of Unified Field Theory, determine the genetic basis of cancer, or concoct a neurological theory of consciousness. There are a thousand fields of scientific endeavour, and very few hit the combination of timeliness and innovation that generate Nobel Prizes (or "the trip to Stockholm," as I overheard one eminent Fellow of the Royal Society describe it). But scientific work is interconnected: like a spider's web, it is sensitive to movement in any part of the structure, and interlinking strands give it its strength. Trilobites, too, connect to large scientific issues: how species are born and die; the "explosive" nature (or not) of the Cambrian; how the biological world we know was engendered; how the ancient continents lay. It is perfectly possible for a researcher to labour for many years, known only to a dozen or so of her colleagues, maintained by love of what she does. Then, by making some connection, she may suddenly be at the cutting edge, lauded by laureates, praised by prize-givers. As in the parables (not to mention the trilobites), those that have eyes to see, they will see. Ruth and Bill Dewel, biology professors from a small East Coast American university, were almost alone in their enthusiasm for tiny, stumpy-legged tardigrades, ubiquitous creatures beneath every clump of moss, that may be close to the ancestral arthropods. The realization that some early Cambrian fossils may relate to them in important features of the head, and the development of molecular methods for determining their evolutionary relationships, propelled these little organisms from peripheral to pivotal. The Dewels' years of patient observation were suddenly germane to all kinds of big questions in the evolution of the most species-rich animals known—the arthropods, in their exuberant variety.

The beauty of the scientific life is that every honest practitioner may add a permanent contribution to the edifice of knowledge. They may be remembered by few of their intellectual successors, but their contribution counts, even if it is anonymous. It is not necessary to be one of the famous few to make a permanent impact. I know that not one person in ten thousand (outside Prague) has heard of the great Bohemian palaeontologist Joachim Barrande, and he is a considerable figure in the trilobite firmament. No matter: his monument survives in the geological maps of his homeland and the very fabric of geological time. The scholar will soon find, if he digs a little further, the name of Barrande attached to a hundred scientific names of important fossil animals. Then he will discover that Barrande, too, made mistakes, most notably a theory for the successive origin of animals from his home area of Bohemia, based upon a mis-correlation of rocks. No matter: the errors are not incorporated into the fabric of the edifice of knowledge. Rather, they provide the stuff for historians to chew over, as they trace the complex progression of ideas from conception to acceptance. In the end, the investigator will discover Joachim Barrande the man, the fussy perfectionist who devoted his life to making known the riches of the Bohemian Palaeozoic rocks to the world, and who named a clam for his housekeeper. Like Marcel Proust, whose neurasthenic obsession crafted one of the greatest and longest of all novels, Barrande devoted his life to his vision, and it was a grand one. Like Proust, Barrande lived in an urban apartment cared for by a strong-minded servant. Science is, at root, another human activity shot through with all the frailties and eccentricities that being human entails. Scientists' life stories are like other life stories, and it is just as much fun gossiping about the details. But, for the edifice of truth, it does not matter whether Barrande was saint or sinner, transvestite or transubstantiationist, so long as he was honest.

It may be access to a kind of immortality, of however unusual a variety, which makes science such an attractive option for intelligent people seeking meaning in their lives. In our secular age, other promises of continued existence *postmortem* have lost their persuasiveness. If moral virtue must too often be its own reward, scientific virtue carries the promise of a reward of permanence, of *making a difference*. At its most blatant the permanence resides in a label, a discovery or an idea married forever to the name of the discoverer: Creutzfeldt-Jakob Disease, Asperger's Syndrome, Heisenberg's Uncertainty Principle, Halley's Comet. In biology or palaeontology the author is for ever tied to a species he described and named for the first time: the trilobites *Illaenus katzeri* Barrande, or *Balnibarbi erugata* Fortey allow us both this small measure of immortality. The reward in other branches of science is similar, if more subtle. Death cannot be cheated, but the discoveries made in one's prime might well outlive bodily decay.

I find that the creative part of the scientific process can be explained rather clearly at the trilobite scale. The pursuit of nuclear physics or physiology employs thousands of scientists. Advances are made which cause regular revolutions in understanding. I am told that nearly all the articles in journals in these fields become obsolete before a decade is up. Workers find it difficult to keep abreast of current discoveries, let alone embrace the fullness of their past. History tends to be jettisoned in favour of keeping up with the pack. At the same time, it is necessary to focus on only a small part of the field, especially because the problems faced tend to be both highly technical and in very competitive areas of advance. Take your eye off the ball for a moment and somebody else will snatch it! The trilobite timescale, by contrast, allows us the leisure to survey the whole of history. We find it easy to connect with Dr. Lhwyd in the seventeenth century, with Linnaeus' contemporaries in the eighteenth century, or with Walcott and Barrande in the nineteenth century. The discoveries of the last hundred years exist in continuity with the past: not seamlessly, for progress is almost always by fits and starts. But Harry Whittington's discoveries about trilobite larvae are clearly built on those of his predecessors, like Sir James Stubblefield or Professor Beecher. We are constantly in touch with our past; and libraries are where we do honour to those who came before us. Our literature never really goes out of date. Though trilobites might lie far to one side of the web of scientific knowledge, they feel the same movements, and respond to the same stimuli, as do scientific disciplines far closer to the centre. This history shows that the past, too, is mutable; and that when new discoveries are made, we re-write historical "facts." The job of the palaeontologist is to reinvent the past. There could be no task more demanding of the scientific imagination.

Some people still believe that science and the arts are somehow opposed—the former dissective, the latter creative. This is the attitude famously encapsulated in the 1950s phrase "The Two Cultures," which was coined by C. P. Snow, novelist and senior civil servant. Snow's synopsis has a far longer pedigree, going back at least to the mystic, poet and artist William Blake, and to those who opposed the experimentalist approach advocated by the Royal Society in England in the eighteenth century, and by other academies in the western world. The artist, it was implied, plumbs greater truths through his imaginative fabrications than the obsessed reductionist, who seeks to explore the secrets of the butterfly by dismantling its wings. The critic's stance is perfectly expressed in this verse by Edgar Allan Poe:

> *Science! true daughter of Old Time thou art!*
> *Who alterest all things with thy peering eyes.*
> *Why preyest thou thus upon the poet's heart,*
> *Vulture, whose wings are dull realities?*

Palaeontology is a "daughter of Old Time," or it is nothing. And throughout this book I have used the image of their own "peering eyes" as the key to understanding the trilobites' world, just as I have consciously linked this image with the observations made by scientists upon the fossil material they wish to bring back to life: eye gazing on eye, unblinking. We peer, therefore we learn. But I regard everything I have described as raw material for the poet. Even the smallest item of scientific revelation can be a matter for joy, and a truth unearthed glitters with the iridescence of a tropical *Papilio* butterfly.

So why are so many people ambivalent towards science and scientists? Several images of the scientist come to mind. First, there is a stereotype of a kind promulgated by TV advertisements, which I might term the "barmy boffin." Bald of pate, but fuzzy above the ears, and with enough facial tics to keep horseflies at bay, the boffin whizzes around in a state of high excitement over his latest discovery—often a gizmo of arcane purpose. The boffin's jacket is a baggy tweed affair with screwdrivers poking out of the breast pocket. The boffin always has thick glasses: for some reason it is *de rigueur* to be short-sighted—and indeed, there is apparently a statistical connection between myopia and intelligence. Boffins always have rather feeble physiques. There is a curious assumption that the development of the brain drains away muscular development. It is as if the brain itself were regarded as a kind of parasite that feeds upon the rest of the body: as the cranium expands so the biceps and pectorals shrink, until I suppose the perfect boffin is a massive brain perched atop spindly limbs, like some kind of stick insect crowned with a calculator. Professor Calculus, in the books about Tintin, the boy detective, was the type example of the boffin: clever as you like, but always vaguely discombobulated, and needing down-to-earth common sense to make a fist of anything. His inventions were always liable to take off with a disastrous life of their own. However, nobody was ever in doubt that Professor Calculus's heart was in the right place. No spoiler of other people's pleasures, his inventions were always somehow linked to something magical—gadgets which might bring the impossible into unpredictable existence. Nowadays, the boffin's equivalent is probably more the intense computer nerd, playing with his machines with the abandoned assurance of the concert pianist. Out of this electronic mastery comes—a beautiful android? a time machine?

But Poe's* scientist is somebody altogether more sinister, a heartless dissecter of innocent animals, perhaps, or a genetic engineer, or a tinkerer with anatomy along the lines of *The Island of Dr. Moreau*. H. G. Wells's story has provided the screenplay for several films; the eponymous Doctor peoples his island with ghastly inter-species grafts. As with much of Wells, what once seemed perverse imagination now seems almost possible, but somehow less sinister. We no longer believe that the implanted heart of a pig might convert the recipient to piggishness. But Wells may have contributed something to the demonization of the scientist by pointing up what happens when technical facility is decoupled from moral responsibility. After all, in the mid-twentieth century we have had examples in the Nazi era which exceeded Wells's most oppressive nightmares. A perpetrator of these aberra-

---

*Edgar Allan Poe himself contributed several scientific ideas on astronomy and biology, which were met with indifference. His prejudice against scientists may not be entirely untinged with personal bitterness.

tions can scarcely be the "vulture whose wings are dull realities" of Poe's poem, for there is a much more active malevolence here than provided by an opportunist scavenger. Both the benign and the intimidating images of the scientist reflect the ambiguity of the role as the layman sees it. On the one hand many people look to the scientist as cure-all, purveyor of the weekly "breakthrough." On the other, the very success of the project, and the arcane language in which much of it is conducted, leads to a feeling of exclusion, of "them" leading us by the nose: we discover a character like Stanley Kubrick's Dr. Strangelove; or the fizzing laboratories that James Bond scuppers for the good of us all.

Trilobites, however, are innocent of all charges. I suspect that the image of the palaeontologist is going to be more Professor Calculus than Dr. Strangelove. Try as I might I cannot devise a scenario whereby trilobite science is appropriated by a totalitarium régime to oppress the people. "Aha, Mr. Bond! You have arrived just in time to witness the triumph of the trilobites, and the end of the human race." I would guess that 80 per cent of scientific endeavour is as innocent of moral implications as the trilobites. Oddly enough, it is precisely because of the harmlessness of such research that it is more difficult to fund; the only science that never has to fight for funding is that with military or medical significance.

So a plague on all the Edgar Allan Poes of our time! The truly dull realities of which they speak are the problems of getting financial support to do work which is not going to yield a commodity marketable within the twelvemonth, which is when the accountants wheel out their electronic abacuses. To determine what is really of value, it is necessary to take an altogether longer view. Consider, for example, the general fascination with dinosaurs, and recall that it was devoted, painstaking scientists who first pieced together these fantastical animals. Sometimes this processs took a decade: digging, preparing, piecing together disparate fragments, lastly putting flesh on the bones. Imagine, if *Tyrannosaurus* had

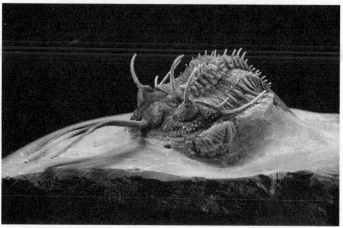

A trident-bearing trilobite from the Devonian of Morocco, as yet unnamed.

remained unknown, how many children's lives would have been impoverished. In the end, careful scientific work was even rewarded financially, if you take into account dinosaur films and books and a hundred less tasteful "spin-offs."

I speculate that my trident-bearing trilobite shown opposite will one day stimulate a moment of wonder in a child, which may convert a waverer to science, enthralled by the marvels that wait to be found. Or even inspire a poet to take flight on some journey of the imagination: to subvert Poe's image, vultures soar effortlessly, and are elegant in flight as an eagle.

Nor can you ever say: now, we know enough. We know a dozen dinosaurs—why do we need to know thirteen? Aren't there enough trilobites in the world? To which I reply: the search is never complete and we can never know what remains hidden behind the next bluff, or inside the next piece of shale. My trident bearer was a dream, a chimera that should not exist; yet it did. The world would have been a poorer place had it remained undiscovered. I anticipate many more such discoveries, facts, if you will, but thrilling ones. Maybe some lucky investigator will discover the limbs of a larval trilobite, so that we may know how they lived as compared with the adult. Can it be that somebody will discover the late Precambrian ancestors of the trilobites, preserved in a state of temporal grace? Will we be able to see the mysteries of this time of innovation explained, as once Walcott surveyed the mysteries of the trilobite limb? These are not Gradgrind facts, "dull realities": they are wings for flights of the imagination. I wish I could live long enough to know, and even if I did, I should never cry "enough!"

It is more difficult to fathom future connections across the web of knowledge, since they depend on advances in a dozen other sciences. My conviction that connections will continue to be made is based upon the fertility of this line of reasoning in the past. Although, as they say of the stock market, "past performance is no guarantee of future profitability," the fact is that these stocks have reliably yielded a return for over a century, and the smart money should still be on them. I can imagine that the physicists will get to work on trilobite optics, and that we will see more clearly how trilobites saw. Their eyes will gaze upon vanished worlds with a pristine clarity. We will learn from molecular studies how all the living relatives of trilobites compare one with another; we'll know what questions we should ask about their anatomical structures. Surely, we will learn more about how trilobites laid down their carapaces; already the electron microscope is probing the details of the minutest crystals. Maybe we will find that trilobite shells record tiny traces of rare elements that act as monitors of dead seas, a mineral equivalent of the pollutants that are picked up in tissues of living creatures, and which are mea-

An entire specimen of the spiny Silurian trilobite *Kettneraspis* from Dudley, Worcestershire, UK. Specimen about 2 cm long. (Photograph courtesy Derek Siveter.)

surable to parts-per-billion thanks to the extraordinary accuracy of modern techonology. We will assuredly get to measure geological time so finely that we can turn the tables on history. Instead of using trilobites as chronometers we will examine their evolutionary changes against a finer-scale timepiece; thus there will be new insights into the evolutionary mechanisms that so intrigued the tragic Rudolf Kaufmann. Trilobites may emerge again as the *Drosophila* flies of the Palaeozoic, the experimental medium for the history of life.

These are dreams of possibilities. Yet I know that there can be nothing better than to pursue such dreams; that the will to know the truth is one of the better parts of human nature; and that trilobites will reward the investigator in a currency more valuable than dollars, and more tangible than fame.

Acknowledgements

My thanks first to Professor Harry Whittington, doyen of trilobitologists, for admitting me into the trilobite trade, and most recently for his generosity in supplying me with photographs. Professors Winfried Haas, Brian Chatterton, Euan Clarkson, Riccardo Levi-Setti, and Drs. Derek Siveter, Bob Owens, Ellis Yochelson and Adrian Rushton, together with the Natural History Museum, kindly supplied numerous other photographs, which have enriched this book. Heather Godwin has been both critic and supporter, and my debt to her is on every page. Robin Cocks read my first draft and made several suggestions which have improved the final work. My wife carefully read the proofs. I thank Claire Mellish for technical support, Pam Hanus and Nicola Webb for translation from the German. Arabella Pike and Michael Fishwick at HarperCollins have been encouraging at all times. Arabella in particular spotted and corrected many small errors, and coped with the complexities of production with unfailing good humour. My fellow commuters on the 8:02 from Henley-on-Thames kept me cheerful on several occasions when I might have succumbed to gloom. This book would not have been possible without the collaboration of my trilobite colleagues scattered all over the world, not enough of whom are mentioned by name.

Suggestions for Further Reading

Kaiser, Reinhard, *Königskinder*, Fischer Taschenbuch Verlag, 1998.

Kowalski, H., *Der Trilobiten*, Goldschneck-Verlag Korb. German book, particularly good on Devonian species.

Levi-Setti, Riccardo, *Trilobites*, University of Chicago Press, 2nd ed., 1984. A photographic atlas, beautifully illustrated with a wide coverage of interesting animals.

Osborne, Roger, *The Deprat Affair*, Pimlico, London, 1999.

Snâjdr M., *Bohemian Trilobites*, National Museum, Prague. Good photographs of many of the famous trilobites of the Bohemian region.

Whittington, H. B., *Trilobites: Fossils Illustrated*, vol. 2, Boydell Press, 1992. 120 plates illustrate many fine trilobites, by the doyen of the field.

Whittington, H. B., and others, 1997. "Treatise on Invertebrate Paleontology," Part O, *Trilobita 1* (revised), University of Kansas Press and Geological Society of America. The standard academic work on the subject.

Index

Note: Page numbers in *italics* refer to illustrations.

Index

Index

Ediacara fauna, 125
Edinburg Limestone, 38–9
  see also limestone
Eldredge, Niles, 162–6
Eliot, T. S., 139
Elrathia, 212, 243
  E. kingi, 239
embryology, 86, 90
Emergence of Animals, The (McMe-
    namin and McMenamin), 141
encrinurids, 185, 241–2
endurance of species, 165
England, 3–10, 55–6
  see also individual place names
enrolment, 55–6, 60–1, 102, 245
environment, 118, 175, 187, 189,
    211–17
  see also climate; habitat
erosion, 8, 137, 194
  see also plate tectonics
Escher, M. C., 87
Estonia, 56
evolution, 159–90
  allopatry, 164–5, 168, 175
  anatomical adaptation, 187,
    211–17
  of arthropods, 234–6
  Cambrian "explosion," 121–45
  cladistics, 132–5, 138, 235n
  creation theory, 159–62, 166
  divergence time, 86, 89–92
  environmental adaptation, 175,
    189
  evidence of, 162–8, 172–5
  and extinction, 181–9
  "failed designs," 129–30, 138
  geographical influences on,
    164–5, 194–211
  homoeomorphy, 208

homology, 124, 187
HOX genes, 85–7, 131, 160
Kaufmann, Rudolf, 166–71
polyphyletic origin, 129
punctuated equilibrium, 162–6
theory of, 25
  see also evolution; genetics
exoskeleton, 41–2, 50, 127, 186
  see also moulting process
extinction, 120–45, 181–9, 218–54
  at end of Devonian period,
    245–6
  mass, 185, 242–3
eyes, 32, 84–119
  Briggs/Fortey tree, 133–5
  calcite, 92–6
  compound, 96, 97, 102
  corneal surface, 101–2
  development of, 87
  evolutionary evidence, 91, 162–6
  field of view, 99–100, 110–16
  genetic origin, 89, 116–17
  and habitat, 113–19
  holochroal, 103n
  lenses, 96–9, 101–4, 106–8
  magnesium, 107–8
  moulting process, 37, 101–2
  phacopid (schizochroal), 103–9,
    186, 244
  progressive illumination, 118–19
  "soft," 94
  spherical aberration, 106–8
  of swimmers, 110–19
  in various species, 74–83
  see also senses

Face of the Earth, The (Suess),
    112–13

275

Index

Index

Index

Index

A NOTE ON THE TYPE

The text of this book was composed in Palatino, a type-
face designed by the noted German typographer Her-
mann Zapf. Named after Giovanni Battista Palatino, a
writing master of Renaissance Italy, Palatino was the first
of Zapf's typefaces to be introduced in America. The first
designs for the face were made in 1948, and the fonts for
the complete face were issued between 1950 and 1952.
Like all Zapf-designed typefaces, Palatino is beautifully
balanced and exceedingly readable.

Composed by North Market Street Graphics,
Lancaster, Pennsylvania
Printed and bound by Quebecor Printing,
Fairfield, Pennsylvania